T0339868

GASIFICATION OF
WASTE MATERIALS

GASIFICATION OF WASTE MATERIALS

Technologies for Generating Energy, Gas, and Chemicals from Municipal Solid Waste, Biomass, Nonrecycled Plastics, Sludges, and Wet Solid Wastes

SIMONA CIUTA

DEMETRA TSIAMIS

MARCO J. CASTALDI

ACADEMIC PRESS

An imprint of Elsevier

Academic Press is an imprint of Elsevier
125 London Wall, London EC2Y 5AS, United Kingdom
525 B Street, Suite 1800, San Diego, CA 92101-4495, United States
50 Hampshire Street, 5th Floor, Cambridge, MA 02139, United States
The Boulevard, Langford Lane, Kidlington, Oxford OX5 1GB, United Kingdom

Notices
Knowledge and best practice in this field are constantly changing. As new research and experience broaden
our understanding, changes in research methods, professional practices, or medical treatment may become
necessary.

Practitioners and researchers must always rely on their own experience and knowledge in evaluating and
using any information, methods, compounds, or experiments described herein. In using such information or
methods they should be mindful of their own safety and the safety of others, including parties for whom they
have a professional responsibility.

To the fullest extent of the law, neither the Publisher nor the authors, contributors, or editors, assume any
liability for any injury and/or damage to persons or property as a matter of products liability, negligence or
otherwise, or from any use or operation of any methods, products, instructions, or ideas contained in the
material herein.

British Library Cataloguing-in-Publication Data
A catalogue record for this book is available from the British Library

Library of Congress Cataloging-in-Publication Data
A catalog record for this book is available from the Library of Congress

ISBN: 978-0-12-812716-2

For Information on all Academic Press publications
visit our website at https://www.elsevier.com/books-and-journals

Working together
to grow libraries in
developing countries

www.elsevier.com • www.bookaid.org

Publisher: Joe Hayton
Acquisition Editor: Raquel Zanol
Editorial Project Manager: Ana Claudia A. Garcia
Production Project Manager: Sruthi Satheesh
Cover Designer: Mark Rogers

Typeset by MPS Limited, Chennai, India

CONTENTS

CHAPTER ONE

Introduction and Background

Contents

The management of municipal solid waste (MSW) in an environmentally sustainable and cost-effective manner is the grand challenge of our time. Yet, it has not received the attention and support that is commensurate with its impact on environmental and human health. In 2015, 2.0 billion tons of waste were generated, while 90% of third-world countries and small-island nations manage their waste through unsanitary landfills and open burning of waste, which is carcinogenic. There are numerous charitable organizations, institutions, nongovernmental, and government agencies focused on solving some of the biggest problems facing humanity. For example, the U.S. Energy Information Agency has quantified that nearly 18% of the world's population does not have access to secure, clean energy [1], the United Nations has identified that nearly 11% of the world's population does not have access to clean water, and the World Hunger Organization has determined that approximately 11% of the global population does not have sufficient food [2]. These percentages translate into an average of 910 million people which is larger than the entire population of the United States and the European Union combined. However, that is only 13% of the entire world population that lives with MSW disposal issues because the *entire* world has a waste generating capacity that must be managed. Yet, there is remarkably little effort compared to other challenges that are funded. The issue of waste management is normally left to regional and municipal agencies to develop an infrastructure focusing on collection. Once collected, the vast majority ($\sim 85\%$) is sent to a landfill or open dump [3].

Waste pollution is an issue that all communities worldwide must control, regardless of geography, culture, or economic standing. Improper waste management can lead to disease, poor standard of living, and environmental decline. One-third of the food generated worldwide is wasted.

Gasification of Waste Materials.
DOI: http://dx.doi.org/10.1016/B978-0-12-812716-2.00001-7

Municipalities and local authorities aim to manage their communities' waste, but they are limited in their efforts when the waste generator does not participate in the established programs. For example, the Department of Sanitation in the City of New York has a 30-year old curbside recycling program to divert recyclables from landfills. However, in 2016, only 16% of recyclable items were actually recycled by waste generators; the remaining 84% of recyclables ended up in a landfill. Although mismanagement of waste may be attributed to human behavior, it is also largely attributed to a lack of education of communities and decision-makers. Communities are uneducated on the details of their waste diversion programs, such as recycling, composting, and reuse, and therefore dispose of many items that could otherwise be diverted from landfills. Furthermore, they are either unaware or uninformed of additional waste-management technologies like anaerobic digestion and thermal treatment that can recover value from their waste. A recent survey conducted by Earth Engineering Center at City College (EEC) | CCNY revealed that 34% of New York residents did not know what energy-from-waste (EfW) was and when they were informed, 88% wanted their waste to be managed incorporating this method [4]. Local authorities aim to assess waste-management techniques and determine a best practices approach for their communities based on available resources and current infrastructure. However, there is a lack of education that spans across all citizenry from the public to the municipal decision-makers.

There are several key players who are involved in addressing the issue of waste pollution—the industry, the decision-makers, and the waste generators. The industry aims to provide a solution by developing technologies and materials that would reduce the negative environmental impact of waste. The decisions-makers strive to assess best waste practices for their community based on available resources and infrastructure. The waste generator is an integral part that needs to be made aware of their impact on society. It must be recognized that waste management is not a one size fits all; it must be customized to the community because waste-management issues vary based on geography, affluence, and social culture. For example, in Ethiopia, a waste-management solution needs to be designed for a high food waste stream and minimal infrastructure, whereas in NYC, solutions need to handle higher concentrations of plastics in the stream and manage multiresidential collection. Therefore, education and dissemination of the available technologies and methodologies are key to helping communities develop waste-management solutions that are best for their community.

Table 1.1 World Bank Data of waste generation per capita by income level

Income level	Waste generation per capita (kg/capita/day)		
	Lower boundary	Upper boundary	Average
High	0.70	14	2.1
Upper middle	0.11	5.5	1.2
Lower middle	0.16	5.3	0.79
Lower	0.09	4.3	0.60

Many studies have been done on estimating MSW generation that range from local municipalities to global data-gathering yet there remains a gap in the technology understanding and deployment available worldwide. The World Bank has one of the most comprehensive studies done on quantifying MSW generation across all countries and provides an informative categorization related to region and socioeconomic standing yet barely discusses technological solutions. Specifically regarding average waste generation as it connects to income level, the World Bank study presents the following data in Table 1.1 where it can be concluded that waste generation increases as a region's wealth increases [3].

This table is misleading in two important ways. First, it sends a message that in lower income regions the amount of waste is about one-third that of high income countries. Second, and more important, it is an incorrect metric. The average amount of waste generated is not the major problem regarding MSW management, it is the total amount of MSW produced and the corollary of how it is managed.

Inspection of the detailed data provided in the World Bank study presents a very different perspective that must be more widely understood and disseminated if any real progress is to be made in the sustainable management of wastes. Fig. 1.1 displays the full data set of waste generation compared to median income level, in $US, for all countries. Overlaid is the average zones that are shown in Table 1.1.

The linear regression line included in the data set does show an increase as the median income rises however the statistical fit is an unacceptable 0.07. The thin bars denote the average median income of the categories used by the study where it can be seen that they are clustered near the lower end of the range. The main observation from this data is there is a large scatter where even at the $100,000 median income mark the generation rate ranges from 2.9 to 6.1 pounds per person per day.

Continuing the analysis on a more statistical basis, Fig. 1.2 shows the averages of generation rate for each category of income as designated by

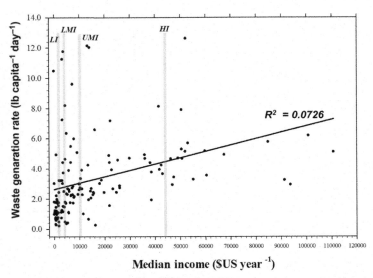

Figure 1.1 Complete data set of World Bank MSW generation compared to median income.

the World Bank with standard deviation bars. Here, it becomes obvious that there is no statistical difference in waste generation rate for the different income categories. Moreover, there is a lower bound that appears to develop near 0.3 pounds per person per day as shown in the data on the left graph in Fig. 1.2.

Taking the data and excluding the incomes below $10,000 $US per year reveals a very interesting result. Shown on the right graph in Fig. 1.2 is the same data presented in Fig. 1.1, yet the abscissa covers an order of magnitude from 10,000 to 110,000 $US per year. Those ranges include countries from Brazil and Croatia to Luxembourg and Norway where it can be seen the waste generation rate appears to be range bound between 1.5 and 5.5 pound per person per day. This data forces us to ask; is there a minimum amount of waste that must be generated to sustain life? It also forces the recognition of how to properly manage the waste to minimize the impact on the environment and human health.

A standard response to the waste generation problem is to develop strategies that will lead to a zero waste utopia or to strive to recycle all waste streams leading to a potential closed–loop of resource usage. Those efforts are unrealistic and establish a false opportunity that further removes the waste generators (i.e., the public) from the real ramifications of their

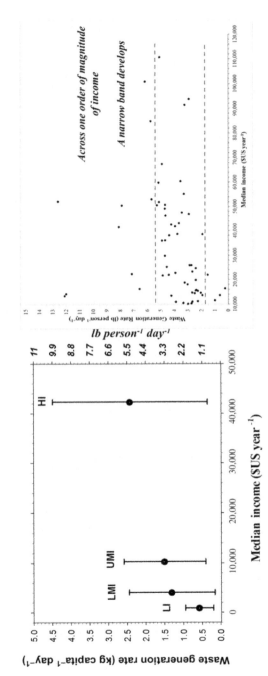

Figure 1.2 Statistical analysis showing the range bound generation rate.

behavior. The goal of a zero waste is a commendable one. The perfect culture would be structured to consume sustainably. However, in the 50 or so years in which the world has been actively promoting the concept of a zero waste society, only a fraction of the solid waste generated worldwide is in fact reused or recycled. Landfilling of waste, particularly municipal waste, continues to dominate the world's waste-management practices. Most people in the developed world are already convinced of merits of environmental protection and would support the goal of zero waste in the abstract. The problem is that zero waste is more of a philosophy than a practical materials management strategy which means it cannot realistically be achieved either technically or culturally.

There is a risk with public perception that all solid waste can be reused or recycled. The short-term risk is the so-called environmentalists who are against any technology and focus on banning materials and asserting "zero-waste" is the solution. Zero-waste does not and will not exist. It is shameful to deny developing areas the benefits of plastics or to saddle them with ideals that cannot be achieved. This ensures that the waste generated continues to create environmental damage and human health hazards. Another short-term risk is the timespan of officials and decision-makers that avoid controversial issues. This leads to paralysis of the agencies and organizations that must take a lead in addressing this issue.

The long-term risks are financial and human resources. For example, at present, a mass-burn waste-to-energy (WTE) facility and a modern designed landfill can manage the garbage at the scale it is generated. However, the expenses of those options preclude poor countries and locales from deploying them. Therefore, other solutions and technologies must be developed or current ones must be adapted to fit. Exactly how this risk will be mitigated is unclear, yet once people are properly informed, they will not divert resources from maintaining the effort.

The sheer volume of generated MSW, coupled with an ingrained culture of disposability, creates a material management system that is inefficient at best, and unsustainable at worst. As local governments struggle with increasing challenges of collection and disposal, radically different approaches are proposed to address the problem. Although most people in the country do not focus on waste management, interested stakeholders seem to be divided into two primary camps. One espouses a cultural and philosophical change to what is called zero waste, holding that the developed world must change consumption patterns to achieve a society in which nothing is wasted and materials are handled in self-contained systems. The other promotes the use of an "integrated" waste-management strategy in which a significant diversion from landfilling via increased recycling/reuse/composting and energy recovery via incineration is considered not only the goal but the only practical solution. Unlike zero waste, the latter strategy has, to a degree, been implemented successfully in many parts of the world.

There are technological solutions available that have been developed during the past decade. These solutions however must be adjusted to fit culturally and economically with different regions as opposed to simple technology transfer. Paramount is the transformation of thinking that waste is a resource and should be positioned to demonstrate its ability to solve problems. A prime example is the inherent energy contained in MSW that if harnessed correctly can have profound positive impacts on a society. In the developed world, most WTE systems combust MSW in well controlled facilities that achieved emissions well below regulations while simultaneously producing about 650 kW h of electricity and nearly 600 kW h of steam for various end users. In developing countries, the infrastructure to fully exploit the energy is not in place; therefore, the use of energy from MSW must be integrated in a different way. Notwithstanding the infrastructure issue, the recognition that at the very minimum the energy must be harvested needs to penetrate the thinking of every region. If a developing country (i.e., 5000 $US per year) develops and integrates a suitable technology to yield the energy in the MSW it can provide a very useful outcome. Specifically if one takes the lower end of waste generation at 2 pounds per person per day, that amount of waste can generate enough energy to irrigate about 45 acre-feet per day. Since it is precisely developing countries that lack access to water or sufficient food or clean energy, the transformation of their waste stream can address a significant portion of the food, energy, water problems they encounter.

Recently, an international conference on waste valorization, WasteEng, engaged discussion among the participants where there was broad participation ranging from experienced to novice participants and from the developed and developing countries. Although the discussion was free format there developed three overarching themes where most of the comments and interactions could be categorized.

One theme centered on the technologies to thermochemically convert biomass materials (agricultural wastes, MSW, etc.) to useful products such as high calorific gases and power and energy. The second general theme touched upon the social and political mindset related to deployment and development of thermochemical processes. The third general category was the interconnectedness of the problem of solid wastes. It is common to all peoples and socioeconomic status and geographical locations confirming the data presented in Fig. 1.2.

The first general theme on technology involved discussion touching upon many facets from suitability of different biomass to costs, development, and the speed and ease of technology transfer and awareness. The leading technology was combustion although a large investigation and acceptance surrounds pyrolysis and gasification systems even though they are much smaller scale and do not have a robust track record. However, there is the potential of these to produce a useable gas that can be distributed to local generation stations or to the population directly. The primary issue with technology development and/or transfer is the cost (price) of the system to procure, install and operate. This is in contrast to the cost (price) of the feedstock (i.e., solid wastes) which has a low price thus making it difficult to finance thermal projects. It was fully recognized that demand will set the price of the entire system but is most impactful on the solid waste feedstock and the energy sector has a large influence on this. For example the price for a certain feedstock in Brazil increased by three times because there was a large demand by the energy sector to obtain a renewable fuel. Within the confines of this category it was recognized that all biomass, and wastes, are not suitable for all processes. Different technologies require different feedstock and processing, however blending of wastes may help to make technologies more universal.

The second broad category or theme merged on was the social and political mindset. Here it was recognized the main issue for people (and governments) was the fundamental need for reliable energy. The developed nations (i.e., European Union and United States) can afford to make decisions based on sustainability goals. However, developing societies are driven by a waste-management goal, and recycling and thermal conversion are based out of necessity not environmental considerations. The developing nations' requirement to secure energy requires they use the most accessible and cheapest, which usually coincide. This is because fossil based fuels are easy to combust whereas MSW needs more attention and specialized systems to extract the energy. It was identified that there is a gap between what is being done/developed/researched in academia and what industry and politicians push forward as solutions. The challenge, or role, is for Academia to work on things relevant and communicate to proper groups such as the government, public, local industry their findings and projections.

Finally, the third general theme addressed the idea that waste (MSW and biomass) is common to all, nobody is without it whether industrialized or emerging countries. Recovering energy and materials from that waste is viewed as an opportunity but technological developments need to continue to enable lower costs and personnel training. This would lead to more deployment and acceptance creating a positive feedback system—i.e., wastes are generated but energy and materials can be recovered thus providing an economic incentive first which may eventually develop into an environmental goal.

A final comment that was made, and received majority agreement, was political and economic stability is paramount to ensuring stable markets. Those stable markets could then develop into deployment of thermal conversion systems recovering energy and materials from waste streams leading to a much more sustainable human endeavor. In most gasification plants, the obtained syngas is subsequently combusted in a steam boiler, to produce electricity in a steam turbine, although this is not the most efficient way of producing electricity from hot combustible gas. Such plants are very similar to combustion plants, but full oxidation is carried out in 2 steps (2 step oxidation): feedstock gasification, followed by syngas combustion [5]. These plants are rather simple to operate, but just like combustion plants only yield electric power and heat (steam, hot water), and may in fact be considered a special sort of combustion plant. Other types of gasification plants ("full" gasification) can provide power and heat and a useable syngas that can be converted to make chemicals (e.g. methanol, ammonia, hydrogen, and liquid fuel) [6]. To produce the chemical compounds, the obtained syngas must be thoroughly cleaned and treated to comply with the specifications of the downstream processes converting the syngas to the desired product. Extensive gas cleaning to remove tar, acids, particles, etc. from the syngas is also necessary, when the syngas is used for high-efficiency electricity (and heat) production in combined cycle turbines, gas engines.

Progress cannot be made at all if a technology does not exist to achieve the desired goals. Development of technologies that convert solid wastes has occurred at a steady pace for the last half-century to attempt to capitalize on some of the drivers listed above. Although several variations have been established, the leading thermochemical technology is combustion, currently referred to as *WTE* or *EfW.* WTE is a proven, highly reliable technology with worldwide recognized equipment manufacturers. It is robust and flexible, as it can operate with various waste feedstocks, ranging in composition and size, at a gate fee typically near 100 €/t. State-of-the-art installations allow stable operation with an availability of at least 8000 h/year. Worldwide, about 250 million tons of waste (mainly MSW) are treated in approximately 2000 plants with an annual capacity of at least 70,000 ton/year.

Gasification is a thermochemical partial oxidation process, using oxygen, air, water, carbon dioxide or mixtures thereof as gasifying agent, that converts a solid or liquid combustible feedstock into syngas, a mixture of mainly CO and H_2. Syngas can be used as a fuel for producing electricity and/or heat, or as building block for the chemical industry. Gasification is not a new idea nor a new technology, since industrial applications for the production of town gas from coke date back to the beginning of the 20th century, and a number of large-scale plants for the production of electricity from coal or heavy oil residues were built in the United States and Europe in the last 30 years. Gasification is often put forward to public administrators as an innovative alternative to the conventional combustion WTE plants [7]. Although much literature is available on gasification of biomass [8,9] and of MSW [10,11], the number of operational waste gasification plants (about 26 operating worldwide corresponding to approximately 1.2 million tons of biomass and waste per year) is two orders of magnitude lower compared to combustion [12]. However this is where the most innovation is observed and reported. Due to the rapid development cycle and technology turnover in the gasification area continuous updates must be provided. There are many texts and peer-reviewed articles that provide gasification information and periodically it must be assembled into a cohesive source.

This book presents EEC|CCNY's global experience on gasification technologies that have been tested and evaluated. It should be recognized that this text is a companion to other texts published [13−19]. The information provided above establishes the full context of MSW from generation to management and associated philosophies. We focus on gasification technologies because these are not as widely deployed and not as well developed as conventional combustion units. Yet, there are tremendous dispersed and small-scale efforts trying to develop gasification technologies for two main reasons. The first is the final product from gasification is heat and a useable gas and the second is gasification is generally more suitable to small-scale operation (~ 100 ton per day). These are import aspects especially for countries and communities that cannot afford state-of-the-art combustion systems or do not have the waste generation rate to support a combustion facility or do not have the infrastructure to best use the electricity and steam.

The text should be viewed as an update of some of the newest gasification technologies that are currently being developed and deployed globally. It should be viewed as a companion to the many other texts that exist on gasification, combustion and material recovery from MSW. The field is so large that it cannot be adequately compiled in any one text because the effort required to assemble such a large body of work would result in much of it being outdated by publication time. Therefore, over the course of one year we have compiled and assembled work done by EEC | CCNY and our associates into the following structure that we think provides an up-to-date view of gasification systems

Chapter 2 focuses on the "Fundamentals of Gasification and Pyrolysis" to provide an overview of gasification systems yet is should not be considered comprehensive. It touches upon pyrolysis and gasification reactions using partial oxidation, steam, and CO_2 as well as hydrogasification. It then provides some information on catalytic reactions and syngas to liquid fuels finishing with plasma gasification, hydrothermal gasification, and supercritical gasification.

Chapter 3 explores "Laboratory/Demonstration-Scale Developments." It is divided into gasification to energy from laboratory-scale plants, gasification to by-products (i.e., chemicals and fuels), innovative gasification technologies, and integration with other systems.

Chapter 4 focuses on "Field-Scale Developments" through a review of the field-scale systems that have been in operation using solid waste streams as input. Information is taken directly from operating data and field measurements. Importantly, connections are made to the laboratory and demonstration-scale highlighting areas of disconnect or alignment.

Chapter 5 provides view on "Emissions" from some gasification systems that have actual measurements. It summarizes published information on emissions from the stack as puts the emissions into context comparing them against other thermal conversion systems using similar feedstock.

Chapter 6 attempts to identify "Critical Development Needs" for further advancement of gasification technologies. Although there are many needs and each are suited for particular operations, we highlight some of the overarching items that are germane to most units. These include regulations and markets, syngas cleaning (tars, sulfur, particulate removal), feedstock characteristics, corrosion issues and auxiliary and parasitic loads, and energy efficiency.

Finally, Chapter 7 provides an "Economic Summary" looking at market drivers to current waste practices. An attempt is made to provide realistic values from analyses and interactions that EEC | CCNY has had with different technology developers.

ACKNOWLEDGMENTS

The author would like to thank numerous students and colleagues who had worked to improve sustainable waste management solutions. Specifically the author acknowledges contributions of this chapter by William F. (Rick) Brandes formerly of the USEPA, Professors Carlo Vandecasteele and Jean-Michel Lavoie, and the participants and organizers of the WasteEng Conference Series. Finally, a grateful acknowledgment is given to many authors who have published some excellent information in this area.

REFERENCES

[1] A. Sieminski, International Energy Outlook, Energy Information Administration (EIA), Washington, D.C., 2014.

[2] J. Parfitt, M. Barthel, S. Macnaughton, Food waste within food supply chains: quantification and potential for change to 2050, Philos. Trans. R. Soc. London, Ser. B: Biol. Sci. 365 (1554) (2010) 3065–3081.

[3] D. Hoornweg, P. Bhada-Tata, What a waste: a global review of solid waste management, Urban Dev. Ser. Knowl. Pap. 15 (2012) 1–98.

[4] C. Cullen, E. Fell, R. Russo, D. Salmon, An integrated waste-to-energy plan for New York City. Capstone Thesis, City College of New York, New York, NY, 2013.

[5] M.R. Lusardi, et al., Technical assessment of the CLEERGAS moving grate-based process for energy generation from municipal solid waste, Waste Manage. Res. 32 (8) (2014) 772–781.

[6] W. Wei, et al., Tar conversion characteristics during catalytic gasification-reforming of municipal solid waste, J. Solid Waste Technol. Manage. 42 (1) (2016) 685–696.

[7] S. Consonni, F. Viganò, Waste gasification vs. conventional Waste-To-Energy: A comparative evaluation of two commercial technologies, Waste Manage. 32 (4) (2012) 653–666.

[8] M. Asadullah, Barriers of commercial power generation using biomass gasification gas: a review, Renewable Sustainable Energy Rev. 29 (2014) 201–215.

[9] P. Basu, Biomass Gasification, Pyrolysis and Torrefaction: Practical Design and Theory, Academic Press, Elsevier: Kidlington, Oxford, 2013.

[10] A. Bosmans, et al., The crucial role of waste-to-energy technologies in enhanced landfill mining: a technology review, J. Cleaner Prod. 55 (2013) 10–23.

[11] L. Tobiasen, et al., Waste to energy—energy recovery of green bin waste: incineration/biogas comparison, Curr. Sustainable/Renewable Energy Rep. 1 (4) (2014) 136–149.

[12] E. Kwon, K.J. Westby, M.J. Castaldi, Transforming municipal solid waste (MSW) into fuel via the gasification/pyrolysis process, 18th Annual North American Waste-to-Energy Conference, American Society of Mechanical Engineers, 2010.

[13] U. Arena, Process and technological aspects of municipal solid waste gasification. A review, Waste Manage. 32 (4) (2012) 625–639.

[14] P. Basu, Biomass Gasification and Pyrolysis: Practical Design and Theory, Academic Press, Elsevier: San Diego, CA, 2010.

[15] A.V. Bridgwater, Advances in Thermochemical Biomass Conversion, Springer Science & Business Media, Netherlands, 2013.

[16] A.V. Bridgwater, D. Boocock, Developments in Thermochemical Biomass Conversion: Volume 1, 2, Springer Science & Business Media, Netherlands, 2013.

[17] N.B. Klinghoffer, M.J. Castaldi, Waste to Energy Conversion Technology, Elsevier, Philadelphia, PA, 2013.

[18] W.R. Niessen, Combustion and Incineration Processes: Applications in Environmental Engineering, CRC Press, New York, NY, 2010.

[19] P.J. Reddy, Energy Recovery from Municipal Solid Waste by Thermal Conversion Technologies, CRC Press, New York, NY, 2016.

CHAPTER TWO

Fundamentals of Gasification and Pyrolysis

Contents

The gasification process consists in the conversion of a solid/liquid organic feedstock into a gas phase, usually called "syngas," a solid phase also known as "char" and a condensable phase called "tar." This conversion requires the addition of an oxidant, such as oxygen, steam or CO_2, at levels below the stoichiometric amount required for full conversion of the carbon contained in the feedstock. In the cases where steam or CO_2 alone or as a mixture are

Gasification of Waste Materials.
DOI: http://dx.doi.org/10.1016/B978-0-12-812716-2.00002-9

the coreactants the gasification process requires a continuous source of heat, usually from an external source. However, the use of oxygen, or air, can transition the process to an exothermic condition depending on the ratio of associated steam and CO_2. The feedstock conversion represents the partial oxidation of the carbon in the material with the gasifying agent, such as air, oxygen, steam, or carbon dioxide providing some heat to the reaction; thus, gasification is generally exothermic and self-sustaining once initiated. The syngas lower heating value ranges from 4 to 13 $MJ/N\,m^3$, depending on the feedstock, the gasification technology, gasifying agent and the operation conditions [1] and can be used for power generation or biofuel production. The following sections present the fundamentals of gasifying and pyrolysis reactions occurring in different media and operating conditions. The complex structure and heterogeneous nature of some commonly used feedstock, such as biomass or municipal solid waste (MSW) currently represent a substantial technological barrier for their efficient conversion into biofuels and chemicals. However, the production of syngas and heat is commonly done with much difficulty. The first step in developing new technologies and improving existing ones is understanding the basic thermodynamics and decomposition mechanisms happening in each stage of gasification. The thermodynamics have been extensively studied and are readily predicted. The decomposition mechanisms however have yet to be determined for a number of reasons ranging from feedstock variability to the complications of gas—solid reactions. The next step, before the design of pilot-scale reactors can be done, is calculating and predicting reaction and conversion sequences using correlation models and software. A great amount of effort was put into understanding and explaining these steps, and the results of some of these studies are presented in this chapter, whereas the discussion of pilot-scale reactors was left for another chapter.

2.1 PYROLYSIS AND GASIFICATION STAGES

The main steps of the gasification process are: drying of the free or bound water present in the feedstock, pyrolysis or volatile matter release, oxidation, and char reduction. An energy change is attributed to each of these stages, which are mainly considered as endothermic processes yet there is data demonstrating pyrolysis can also be exothermic under certain conditions [2]. The oxidation step always generates some heat, depending on how much free oxygen is present, and is considered an exothermic stage.

2.1.1 Drying

The drying stage of any thermochemical process consists of the evaporation of the moisture contained in the feedstock. As drying is an endothermic process, the heat required for evaporation is proportional to the moisture content of the feedstock.

When feedstock with high moisture content is processed, depending on the gasification technology, drying is imperative. Most of the gasification technologies require the feedstock moisture to be below a certain level, anywhere from less than 10% for the Lurgi technology to less than 60% for the Foster Wheeler gasification technologies [3].

Drying is considered beneficial because it permits smaller dimensions of the reactor and enables a higher temperature to be achieved more readily. High moisture content in the in-feed material would decrease the gasifier temperature shifting the reaction equilibrium toward higher methane concentration and lower hydrogen content in the product gas.

The drier adjacent equipment, such as fans, pumps, chillers, mechanical drives increase even more the power consumption of the overall process. A solution to decrease the energy demand might be using some of the waste heat from other processes if available, or choosing a gasification technology more fitted for wet in-feed material. Although many well-developed waste heat recovery technologies are available today, most of them face technical and economic barriers. Technology developers will only consider investing in such waste recovery systems only if returns are significant and risks are minimal.

2.1.2 Pyrolysis

Pyrolysis is a complex process influenced by several parameters (e.g., waste composition, moisture content, temperature) that directly affect the yields and characteristics of the products obtained. The pyrolytic decomposition is an endothermic conversion step of solid organic matter into gaseous compounds and solid residues with high carbon content. The chemical degradation reactions occurring within the macromolecules of the feedstock do not require the presence of a reactant and the decomposition is mainly heat driven. Part of volatiles released during pyrolysis may condense on colder surfaces and form liquid products called "tars," whereas the permanent gaseous species remain in gas phase as: CO_2, H_2, CO, CH_4, and light hydrocarbons.

The release of moisture occurs at the early stages of heating up the feedstock up to 150−200°C, after which pyrolytic decomposition starts and finishes at temperatures around 500–700°C. At temperatures below 350°C, the main reactions are depolymerization as well as dehydration, decarbonylation, and decarboxylation releasing water, carbon monoxide, and carbon dioxide, respectively [4]. At temperatures above 350°C, more aromatic products are released.

2.1.2.1 Pyrolysis of biomass

Pyrolysis of biomass is a complex process, which has been studied extensively in the past decades. It involves many physical and chemical processes such as drying, heat transfer, solid–gas reactions, endothermic and exothermic energy change, and continuous variation in the material properties.

Chemical bonds of biomass break during pyrolysis releasing volatiles, during the primary reactions of pyrolysis. Volatile matter consists of permanent gases and some condensable species called tars. During this step, also part of char is formed. During the secondary reactions, a portion of the tar decomposes to form secondary tars and gas and polymerizes to secondary char. Secondary tar decomposition could occur homogeneously in the vapor phase or heterogeneously on the surface of the solid.

Secondary tar cracking reactions were found to be related to residence time of volatiles at temperatures between 300°C and 700°C. Increasing the residence time of tar within the solid–gas interface leads to tar cracking, release in hydrogen, methane and forming of solid carbon [5].

An example of relevant secondary reactions shown in Eq. (2.1) is the formation of aromatic radicals [6]. Other examples are shown below in the "tars" section of this chapter.

$$R\text{-}CH_2CH_2\text{-}R' = R\text{-}CH_2 + R'\text{-}CH_2 \tag{2.1}$$

The partition between each pyrolysis by-product (i.e., solid, liquid, and gaseous) depends on the operating conditions such as: reactor type, heating rate, residence time, particle size, pressure, and maximum temperature [4], but also on the amount of cellulose, hemicellulose and lignin within the biomass.

The three main components of lignocellulosic biomass, cellulose, hemicellulose, and lignin, decompose with different kinetics and at different temperatures. Between 200°C and 300°C, hemicellulose undergoes thermal decomposition, whereas in the temperature range 325–375°C, degradation of cellulose is dominant. Lignin decomposition occurs mainly at temperatures above 375°C [7].

One poorly understood aspect of pyrolysis is the actual energy change that occurs during the process. This has applicability with respect to the overall energy balance in a landfill and specifically the energy generation (or consumption) through the pyrolysis process. The heat of reaction has a significant influence on thermal conversion routes, and understanding the effect of the reaction heat is important in modeling thermochemical processes ultimately leading to predictive capability. Unfortunately, reports of the thermal effects of pyrolysis reactions vary widely, ranging from exothermic to endothermic under similar conditions. Furthermore, the exothermic−endothermic variation is observed over a range of conditions [8−11]. For example, the thermal degradation of cellulose is endothermic, whereas char formation reaction is exothermic. However, these processes occur simultaneously during thermal decomposition, and leading to a composite result. In addition, subsequent reactions involving gaseous intermediates are difficult to quantify and may provide additional endothermic conditions due to cracking-type reactions or exothermic conditions related to condensation reactions.

Experimental and modeling studies in the field of biomass pyrolysis [12−14] have focused on explaining the pyrolysis mechanisms to transform feedstock into final products, but fundamental aspects of the mechanisms are not fully understood. Many studies have focused on temperature profiles evolving during the process [15,16] without correlation with gaseous and liquid reaction intermediates and end products that are needed to understand the reaction mechanisms.

Understanding this has been one of the topics of research at the Combustion and Catalysis Laboratory, with several publications describing this in more detail [17,18]. This has real-scale applicability on determining the overall energy balance and the energy consumption through the process. The heat of reaction has a significant influence on decomposition mechanisms and understanding the reaction heat is important in modeling thermochemical processes and predicting the reaction sequences.

Several types of models are available that predict cellulosic decomposition pathways. They vary from oversimplified (i.e., a one to two step reaction sequence that globally relates solid reactants to gaseous and liquid products) to detailed elementary step reactions [19−24]. These models have advantages and drawbacks. Elementary models have information that is very precise and detailed, but they involve a large number of stable and intermediate chemical species and are difficult to parameterize and can be impractical engineering design. The simplified models suffer from inadequate representation of products of the reactions. The simplified models are limited mainly to kinetics of the mass loss of solid, but not the chemical composition of the products. Therefore, the energy changes and heat released (or consumed) are not well quantified.

2.1.3 Oxidation

The char-oxidation (2.2) and partial-oxidation (2.3) reactions are slow in the low temperature regime compared to other reactions described below and consequently do not represent a rate controlling factor in the overall gasification process [25].

$$C + O_2 \leftrightarrow CO_2 \quad H = -393.5 \text{ kJ/mole} \tag{2.2}$$

$$C + 1/2O_2 \leftrightarrow CO \quad H = -110.5 \text{ kJ/mole} \tag{2.3}$$

When air is used as gasifying medium, the oxygen is consumed in the reactor, whereas the carbon dioxide concentration increases proportionally. Oxygen burns a portion of the carbon in the feedstock until all free carbon, elemental carbon in an uncombined state, is exhausted and penetrates the material surface only to a small extent, because O_2 more easily reacts at the surface with the formed carbon monoxide and hydrogen gases. The oxidation zone has the highest temperature due to the exothermic nature of the reactions and provides the energy required for further steps.

2.1.4 Char reduction

The reduction step involves the products from the previous stages, reactions between the gas components and the char, forming the final syngas. The main reactions occurring during the reduction stage are as follows:

$$C + CO_2 \leftrightarrow 2CO \quad H = 172.4 \text{ kJ/mole Boudouard reaction} \tag{2.4}$$

$$C + H_2O \leftrightarrow CO + H_2 \quad H = 131.3 \text{ kJ/mole reforming of the char} \tag{2.5}$$

$$CO + H_2O \leftrightarrow CO_2 + H_2 \quad H = -41.1 \text{ kJ/mole water gas shift reaction} \tag{2.6}$$

$$C + 2H_2 \leftrightarrow CH_4 \quad H = -74.8 \text{ kJ/mole methanation} \tag{2.7}$$

$$CH_4 + H_2O \leftrightarrow CO + 2H_2 \quad H = 206.1 \text{ kJ/mole steam reforming of methane} \tag{2.8}$$

$$CH_4 + 2H_2O \leftrightarrow CO_2 + 4H_2 \quad H = 165 \text{ kJ/mol steam reforming of methane} \tag{2.9}$$

The contribution of Boudouard and char reforming reaction to this step, make the gasification stage overall endothermic [1], although reactions (2.2) and (2.3) are exothermic.

Reactions $(2.4)-(2.7)$ are chemical equilibrium reactions and therefore products and reactants can coexist and maintain their concentration ratios as defined by the laws of thermodynamic equilibrium.

The conditions of equilibrium shift toward the formation of products at higher temperatures for reactions (2.4) and (2.5) and for reactions (2.6) and (2.7) at lower temperatures [1]. Therefore, controlling the temperatures, at which the reduction stage takes place, has a significant impact on the resulting syngas characteristics. Higher temperatures reduce the formation of tars and yield less solid residue at the end of the process, but favor ash sintering and yield a lower energy content syngas [1].

As temperature increases, methane concentration would decrease following the steam reforming endothermic reactions, whereas hydrogen concentration would increase following reactions (2.5), (2.8), and (2.9). CO concentration increases by reactions (2.4), (2.5), and (2.8) that are more dominant than exothermic reaction (2.3). Although endothermic reaction (2.9) releases CO_2, the CO_2 concentration decreases with temperature increase because reaction (2.4) is more dominant, changing the equilibrium toward the right and producing more CO as temperature increases [26].

Therefore, the gasification step occurs when the solid products of pyrolysis decomposition are transformed into a mixture of partially still oxidizable gaseous products [4]. Based on these findings, gasification technology has evolved to offer multiple solutions depending on the desired end product. Typical gasification processes take place in the $800-1000°C$ temperature range, whereas the technology which uses only oxygen as gasifying agent occurs in the range of $500-1600°C$ [1].

2.2 PYROLYSIS AND GASIFICATION BY-PRODUCTS

As the present chapter discusses extensively the main gasification by-product, the syngas, this section focuses only on the solid and condensable by-products of pyrolysis and gasification stages.

2.2.1 Tars

Tar formation is an important step in pyrolysis, especially for liquid fuels production from biomass feedstock. Depending on the operating conditions, bio-oil yield could vary from 25 to 75 wt% [25] and would require an expensive treatment due to the many oxygenated compounds containing and low energy value.

Tar formation can be problematic for downstream applications such as clogging of lines due to condensation. Tars are present in the gas fraction and depending on the technology could vary between 1 and 180 g/N m^3 [27].

Tars can be reduced by catalytically active compounds within the feedstock or added to the process and/or by increased operating temperatures and/or residence times. Steam and oxygen added downstream of the gasification stage can also reduce tars, as explained below. Tar formation starts at temperatures between 200°C and 300°C and then rapidly increases at 300−400°C.

Cellulose and hemicellulose decompose and yield primary tars containing oxygenated organic compounds such as alcohols, carbon-acids, ketones, and aldehydes, whereas the lignin pyrolysis yields mostly aromatic compounds, bi- or trifunctional substituted phenols such as cresol, xylenol [1].

As temperature increases, above 500°C and in the presence of an oxidant such as oxygen, air or steam, the primary tars usually follow dehydration, decarboxylation (2.10) and decarbonylation (2.11) reactions which yield more gas and secondary tars. Secondary tars are alkylated mono- and diaromatics including hetero-aromatics such as pyridine, furan, dioxin, and thiophene [1].

$$O = R\text{-}OH = R\text{-}H + O = C = O \qquad (2.10)$$

$$R\text{-}CHO = R\text{-}H + CO \qquad (2.11)$$

Secondary tar decomposition can occur within the biomass (intraparticle) or outside (extraparticle) the bed of material [28] and the reactions include homogeneous vapor phase cracking and heterogeneous conversion on the char surface. Homogeneous reactions mainly occur above 600−700°C.

Above 800°C, tertiary tars may form and mainly consist of aromatic and poly-nuclear-aromatic-hydrocarbons (PAH) such as benzene, naphthalene, phenantherene, pyrene, and benzopyrene [1]. These tars result from the recombination (2.12), demethylation (2.13) and rearrangement of secondary tars and only appear after primary tars have fully converted into secondary tars [1,5].

$$R\text{-}H + R'\text{-}H = R\text{-}R' + H_2 \qquad (2.12)$$

$$R\text{-}CH_3 + R\text{-}H = R\text{-}R' + CH_4 \qquad (2.13)$$

Primary tars such as levoglucosan, hydroxyacetaldehyde, furfurals, and methoxyphenols are mainly derived from cellulose, hemicellulose or lignin. Secondary tars are represented by phenols and olefins, whereas tertiary tars are divided into: alkalized tertiary products (methylacenaphthylene, methylnaphthalene, toluene, and indene) and simplified tertiary products (AHs/PAHs without substituent) [28]. More representative tar formation with temperature is seen in Table 2.1.

A simple reaction mechanism, adapted from Molino et al. [1] is presented in the following equations. The C_nH_m represents the generic tars and $C_{n-x}H_{m-y}$ represents the hydrocarbons with molecular weight lower than of tars.

$$\alpha C_nH_m \rightarrow \beta C_{n-x}H_{m-y} + \delta H_2 \quad \text{endothermic, tar cracking} \quad (2.14)$$

$$C_nH_m \rightarrow nC + (x/2)H_2 \quad \text{endothermic, carbon formation} \quad (2.15)$$

$$C_nH_m + nH_2O \rightarrow (n + x/2)H_2 + nCO \quad \text{endothermic, steam reforming} \quad (2.16)$$

$$C_nH_m + nCO_2 \rightarrow (x/2)H_2 + 2nCO \quad \text{endothermic, dry reforming} \quad (2.17)$$

The operating conditions will have an effect on the tar production, for example increasing the temperature would decrease the amounts of components like phenol, pyridine, and cresol in the gas [29]. These heterocyclic components are least stable and easily break down at temperature between 750°C and 850°C.

Temperature causes tars to react under inert conditions, also called thermal cracking or with some of the gaseous components in the syngas, such as H_2, H_2O, and CO_2. High temperatures, 1200°C or above [28]

Table 2.1 Tar formation depending on operating temperature

Temperature (°C)	Tar species
400	Mixed oxygenates
500	Phenolic ethers
600	Alkyl phenolics
700	Heterocyclic ethers
800	PAH
900	Large PAH

Adapted from A. Molino, S. Chianese, and D. Musmarra, "Biomass gasification technology: The state of the art overview," *J. Energy Chem.*, 25, 1, pp. 10–25, 2016.

are typically required for thermal cracking in order to obtain a tar-free syngas which can safely be used in downstream gas engines, gas turbines or synthesis processes. The atmosphere and the hydrogen pressure also influence the rate of thermal cracking. It was found that H_2O and/or CO_2 gasification agents increase the tar decomposition rate, whereas the H_2 reduces it [28].

Operating at temperatures below 850°C, for lower energy consumption, would increase the amount of char and tar yielded. The catalytic gasification has been proven to be successful for tar reducing even at lower temperatures.

The most researched catalysts are Ni-based and dolomites. These types of catalysts are added into the gasifier or into a reactor placed in the path of the syngas.

Nickel-based catalysts present higher tar conversion rates, but the reforming reactions occurring are endothermic and significantly decrease the efficiency of the process. Also high costs and severe risk of deactivation from sulfur chemisorption and carbon deposition [30], motivated researchers to look into other options [29]. Importantly, when catalysts are added to the system in this manner, the recovery of the catalyst is impossible. A better solution is to position the catalyst in the downstream section [31].

When minerals such as dolomite and olivine are used as tar cracking catalysts, only tar molecules are reformed, whereas low hydrocarbons concentrations remain the same [30].

2.2.2 Char

The gasification of the remaining char is an important step in the overall process, being a controlling step which determines the final conversion achieved in the process. The reactivity of the char has proved to be an important parameter in reaction mechanisms occurring, thus understanding this step is essential for design of gasification reactors. The nature and the thermal history of the char, its pore structure, and feedstock chemical composition are considered to be mainly responsible for the differences in results [32]. In addition, the mineral material in the feedstock was found to catalyze gasification reactions and influence the process [32]. An important parameter is the catalytic compounds in or added to the feedstock, which would accelerate some reactions and change the final composition of the end products. Char was also found to increase the reactivity and catalyze the water gas shift reaction, by generating more H_2 [4].

The Combustion and Catalysis group in collaboration with University of Toulouse, France looked at the catalytic activity of char from biomass gasification [31] and found that not only inorganics play a significant role in methane cracking reactions, but also the char carbon content. The research carried out at Combustion and Catalysis Laboratory (CCL) focused on understanding the properties of char which mostly influence its catalytic activity. Moreover, increasing the oxygenated functional groups of chars obtained by biomass gasification was investigated [33] in relation with their reactivity.

2.2.2.1 Activated carbon

One other important use of char is preparation of activated carbon. When using coals as feedstock, the material is oxidized and then carbonized into char. The activation of char could follow two different pathways: the chemical activation or the physical path using gasification. The purpose of the activation is to increase the surface area of the char, which after oxidation and carbonization has only fine micropores.

The gasification stages are referred to as degrees of burn-off and produce activated carbons with different size distribution. Some research groups found that activated carbons produced from coals by physical activation using CO_2 gasification would result in wider microporosity than steam gasification [34]. However, an increase in burn-off results in greater meso- and macroporosity and decrease in the number of small micropores [34]. CO_2 gasification generates an increase in macropores up to about 50% burn-off, followed by a considerable decrease at higher degrees of burn-off, whereas a continuous increase occurs when steam is used as gasifying agent [34].

2.3 PROCESS CONDITIONS

2.3.1 Gasification agents

Air, oxygen, steam, or carbon dioxide as well as combinations of these media are commonly used as gasification agents and heavily influence the product gas composition.

Using air as gasifying agent yields a product gas with heating values between 4 and 6 MJ/m^3 and high nitrogen concentrations, whereas oxygen and steam gasification would produce a higher heating value gas (10−20 MJ/m^3) [3] with relatively high CO and H$_2$ concentrations.

Steam gasification is known to be a highly endothermic process, where additional heat is supplied via partial oxidation of the feedstock, through the gasifier walls or via preheated steam.

When using steam as gasifying agent, the resulting syngas would be nitrogen free, low in tars and high in hydrogen content. The steam gasification would yield the same permanent gas components (CO$_2$, CH$_4$, and light hydrocarbons) and some higher hydrocarbons (C$_n$H$_m$), which would react according to reactions (2.18) and (2.19) to yield CO and H$_2$.

$$C_nH_m + nH_2O \rightarrow nCO + (n + m/2)H_2 \qquad (2.18)$$

$$C_nH_m + nCO_2 \rightarrow nCO + (m/2)H_2 \qquad (2.19)$$

Reactions (2.18) and (2.19) are also known as steam and dry reforming, respectively and reaction (2.19) is seen when using CO$_2$ as gasifying agent. The gas composition will be influenced by the CO-shift reaction (2.20) [35].

$$CO + H_2O \leftrightarrow CO_2 + H_2 \qquad (2.20)$$

In addition to the homogeneous gas−gas reactions (2.18)−(2.20), heterogeneous solid−gas reactions occur when the gasification agent reacts with the char formed (reactions (2.21) and (2.22)). The kinetics of the latter are considered to be much slower than the gas−gas reactions [35].

$$C + H_2O \leftrightarrow CO + H_2 \qquad (2.21)$$

$$C + CO_2 \leftrightarrow 2CO \qquad (2.22)$$

As seen above, initially carbon monoxide and hydrogen are formed during gasification and the final ratio CO/H$_2$ is influenced by the gasifying agent, temperature and pressure of within the process. This CO/H$_2$ ratio for different gasifying agents is presented in Table 2.2, as adapted Pollex et al. [4]. Carbon dioxide as gasification agent favors the production of CO, while using steam would yield more H$_2$ in the final syngas.

The Combustion and Catalysis Laboratory as well as a number of other research laboratories have studied extensively gasification under CO$_2$ and steam conditions [36] and concluded that:
- The use of steam as gasification agent enhances the heating value of the product gas, compared to pyrolysis in N$_2$ atmosphere

Table 2.2 Influence of gasification agent on the CO/H_2 ratio of the product gas

Gasification agent	General equation	CO/H_2 ratio
Carbon dioxide	$C_mH_n + m\,CO_2 \leftrightarrow 2m\,CO + 0.5n\,H_2$	2.4:1
Air/oxygen	$C_mH_n + 0.5m\,O_2 \leftrightarrow m\,CO + 0.5n\,H_2$	1.2:1
Water	$C_mH_n + m\,H_2O \leftrightarrow m\,CO + (0.5n + m)\,H_2$	1:1.8

Adapted from A. Pollex, A. Ortwein, and M. Kaltschmitt, "Thermo-chemical conversion of solid biofuels," *Biomass Convers. Biorefinery*, 2, 1, pp. 21−39, 2012.

- Under CO_2-N_2 atmosphere, lower H_2 and hydrocarbons were released compared to pyrolysis, but higher CO and lower CO_2 values were measured
- The CO_2 increased the final mass conversion of solid carbon feedstock (biomass, coal, and MSW) gasification and reduced the tar content by expediting cracking [37]

2.3.2 Heating rate

The heating rate also influences the pyrolysis and gasification by-product yields. Higher heating rates would decrease the char yield and increase the gas product. Research conducted on biomass feedstock [27] showed that less anhydrocellulose, which gives high char yields, is formed at higher heating rates. Rapid heating would decrease the biomass residence time at temperatures below 300°C, and therefore, the cellulose dehydration into andhydrocellulose does not occur, favoring cellulose depolymerization and volatiles formation [27].

2.3.3 Heat requirements

As the gasification process is overall endothermic, heat has to be supplied to the reactor by partial combustion of the feedstock, also called autothermic gasification or by external heating using a heat exchanger or heat carrier, also called allothermic gasification [35].

The autothermic process uses an oxidant such as air or oxygen to combust a part of the feedstock and generates the amount of heat required by the endothermic reactions. The reactor temperature is controlled by the oxidant feed rate.

The allothermic processes, also known as indirectly heated gasifiers could use a bed of hot particles, such as sand which is fluidizes using steam or air. Other technologies use superheated steam to provide the necessary heat.

2.4 LIQUID FUELS

Liquid fuels production occupies a separate section in this textbook due to the attention it has received in the past decades and being most suitable to substitute crude oil as fuel for transportation. Most liquid production is aimed at biofuel synthesis especially for common fuels such as diesel. For example, bioethanol production has tripled from 2001 to 2010, whereas the biodiesel production has increased by an order of magnitude [1]. However, high production costs are still associated with biofuels. The figure below shows the most common pathways of syngas use for liquid fuels production (Fig. 2.1).

The feedstock for biofuel production is represented mostly by biomass such as rapeseed for biodiesel production and wheat and barley grains for bioethanol production [38]. Scientists have also explored biofuels production from different types of biomass such as lignocellulosic materials and algae or plastics separated from the MSW stream, and the predictions are that these types of feedstock are going to increase in the future.

Syngas was used to convert catalytically into methanol and more recently into ethanol. The cost of ethanol from gasification is estimated at $2 per gallon [39], but with continuously changing markets and slow increase in price of petroleum, ethanol could become competitive very soon.

2.4.1 Fisher–Tropsch synthesis

The conventional widely used method to produce liquid fuels involves use of catalysts to refine gases into different fuels. The most common method is the Fisher–Tropsch (FT) synthesis using transition metal-based

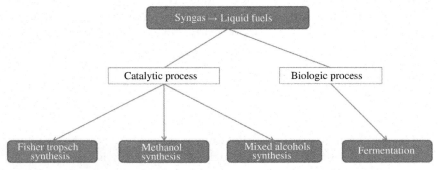

Figure 2.1 Syngas-to-liquid-fuels processes.

catalysts to produce diesel and gasoline. After syngas cleaning, the gas passes through the FT reactor where the reaction between H_2 and CO produces hydrocarbons of varying molecular weight.

The main reaction occurring in the FT process is [1]:

$$(2n + 1)H_2 + nCO \rightarrow CnH_{(2n+2)} + nH_2O \qquad (2.23)$$

The reaction is usually performed at pressures of 20−40 bar and temperature ranges of 200−250°C or 300−350°C [40]. For an optimal FT synthesis, the H_2/CO ratio is required to be around 2 [1], and for syngas this number can only be achieved by using catalysts. The most commonly used catalysts are iron or cobalt based. When using iron catalyst, the H_2/CO ratio does not need to be as high and could be around 0.5−0.7 if catalyst promoters are used [3,40]. This is because iron has intrinsic water−gas-shift activity and the WGS reaction is a standard method to adjust the required ratio, by reacting part of the CO with steam and form more H_2. As cobalt does not have WGS activity, the required ratio is above 2 and the main by-product of the FT synthesis will be water [3].

Iron-based catalysts are typically used in the 300−350°C temperature range and olefins for a lighter gasoline are produced, whereas reactions occurring in the lower temperature range and in the presence of cobalt would generate waxy, long chained products that can be cracked to diesel [40].

Among the other requirements for FT synthesis, we remind:
- Low sulfur content, which is known to reduce the catalyst activity and its lifetime
- Tars are desired to be below 10 ppb and are considered contaminants for the synthesis process as they easily condense onto surfaces, causing the same effect of catalysts lifetime reduction. Tars were found to cause more problems in fixed bed catalysts, whereas slurry bed reactors seem to tolerate more tar contamination [40]
- Nitrogen and methane and other gases which are not reacting during the synthesis are desired to be low in volume, due to the size and cost of the equipment needed when higher volume rates are treated.

2.4.2 Pyrolysis to liquid fuels (bio-oil)

Liquid fraction from pyrolysis can be converted into a fuel similar to petroleum-derived fuels, by reducing its tar and oxygen content. The most common methods used are: catalytic cracking, carbon−coke elimination and hydrogen addition [38].

Most studies about biomass into fuels and chemicals conversion focus on modified zeolites as catalyst which increases the aromatics production.

One known problem of fuels derived from biomass is the high oxygen content around $45-50$ wt% in the final product. Catalytic hydro-deoxygenation is one of the technologies to decrease the oxygen content from bio-oil by reacting with hydrogen to form water and saturated $C-C$ bonds. This method transforms the bio-oil into a more stable and high energy density fuel, comparable to petroleum derivate feedstock.

Rhodium-based catalyst are very effective for syngas to bioethanol conversion, because their C^{2+} alcohols high selectivity [1]. Due of their high costs, studies have looked into more affordable substitutes such as copper-based catalysts. When copper-based catalysts are used, promoters such as alkali transition metals and their oxides [1] are required to maximize the desired end product, ethanol. The main reaction occurring is as follows:

$$nCO + 2nH_2 \rightarrow C_nH_{2n} + OH + (n-1)H_2O \qquad (2.24)$$

Temperature influences the reaction and the end product. C2$-$C3 alcohol high production is achieved in the $833-858°C$ temperature range [1].

2.4.3 Methanol

Production of methanol from syngas is another catalyst driven process, generally occurring in the presence of cooper$-$ or nickel$-$alumina-based catalyst.

Methanol synthesis occurs at temperatures around $220-300°C$ and pressures of $50-100$ bar. The methanol synthesis reaction is equilibrium controlled and involves reacting CO, H_2 and steam in the presence of copper$-$zinc oxide catalysts and small amounts of CO_2. The reaction path for methanol synthesis [3] (2.27) follows the water$-$gas-shift (2.25) and the hydrogenation of carbon dioxide (2.26) reactions:

$$CO + H_2O = H_2 + CO_2 \qquad (2.25)$$

$$3H_2 + CO_2 = CH_3OH + H_2O \qquad (2.26)$$

$$2H_2 + CO = CH_3OH \qquad (2.27)$$

Direct hydrogenation of CO is another pathway for methanol synthesis, but at much slower rates (2.28):

$$2H_2 + CO = CH_3OH \qquad (2.28)$$

The methanol synthesis process involves simple chemical reactions and is less complex than the FT or mixed alcohol processes.

Syngas requirements for methanol synthesis process are [3,40]:

- H_2/CO ratio of at least 2 when using alumina supported catalysts
- CO_2/CO ratio of 0.6 to avoid catalyst deactivation
- Low concentration of nonreactive gases to prevent build up within the synthesis loop
- Low concentrations of CH_4 and C^{2+} to limit the need for steam reforming
- Removal of tars, which can lead to catalyst deactivation
- Eluding of alkali metals which usually promote other reactions, such as FT

2.4.4 Mixed alcohols

Mixed-alcohols synthesis is another process used to obtain alcohols (methanol, ethanol, propanol, butanol) from syngas and is similar to FT and methanol synthesis. The process also known as Higher Alcohol Synthesis uses catalysts and added alkali metals to promote the mixed alcohols reaction [40].

2.4.5 Dimethyl ether

Bio-dimethyl ether can be obtained from methanol or directly from syngas using bifunctional catalyst. Several studies [1] have looked into copper-ZSM-5 zeolite, hybrid copper−alumina-based catalysts, lanthanum oxide, or niobium−alumina for DME production.

2.4.6 Fermentation

Another pathway to produce liquid fuels uses enzymes or microorganism to refine syngas [39]. This method is known as syngas fermentation and is led by acetogenic bacteria, which transform CO and/or CO_2 and H_2 into acetyl-CoA in the presence of CP dehydrogenase [1].

Table 2.3, adapted from Ciferno and Marano [3], summarizes the requirements of gasification gas product for liquid fuels and fuel gas production.

If liquid fuels produced from syngas by catalytic conversion are the desired end product, cost for additional increase of H_2 should be considered. As many of the gasification technologies mostly produce a CO rich

Table 2.3 Syngas characteristics for different applications

Characteristics	Synthetic fuels FT & diesel	Methanol	Hydrogen	Fuel gas	
				Boiler	Turbine
H_2/CO	0.6	2	High	Not important	Not important
CO_2	Low	Low	Not important	Not critical	Not critical
N_2	Low	Low	Low	Not critical	Not critical
H_2O	Low	Low	High	Low	Tolerable
Contaminants	Low sulfur Low PM	Low S Low PM	Low sulfur Low PM	Small	Low PM Low metals
Heating value	Not important	Not important	Not important	High	High
Pressure, bar	20–30	50 (liquid phase) 140 (vapor phase)	28	Low	400
Temperature, °C	200–400	100–200	100–200	250	500–600

Adapted from J.P. Ciferno and J.J. Marano, "Benchmarking biomass gasification technologies for fuels, chemicals and hydrogen production," *US Dep. Energy. Natl. Energy,* June, p. 58, 2002.

syngas, a degree of water gas shift reaction would be required to adjust the H_2/CO ratio.

Also for these type of synthesis processes a high volume of inert gases would increase the cost of equipment, therefore oxygen enriched gasification or steam gasification are the most preferable technologies.

2.5 PLASMA GASIFICATION TECHNOLOGY

The plasma gasification technology is applied to carbon-based materials in a controlled environment with the energy flow coming from plasma torches. The plasma is an ionized gas stream at temperatures up to 10,000°C, which is obtained from the application of an electrical discharge. The plasma state is maintained between two electrodes of the torch by electrical and mechanical stabilization that are built into the plasma torch hardware [41].

The material fed into the process is similar to the other gasification technologies discussed above, and in the presence of air, oxygen or steam as gasifying agents the gasification reactions occur. In the absence of any oxidizing agent, the plasma process is similar to high temperature pyrolysis [1].

The plasma column is generated between two attachment points, one at the solid–gas interface at the cathode electrode and the other at the gas—solid interface at the anode electrode [41]. An insulator is placed in between the electrodes. Because of the high temperatures reached at the attachment points, part of the electrode material usually vaporizes and water cooling is used to minimize it and increase the electrode lifetime.

The plasma technology can be directly applied to the feedstock, with the purpose of thermal destruction or applied to the gas produced from the gasification process. The first type of process accepts a wide variation in feedstock quality, moisture and flow rate. The second process has more of a cleaning and removal of tars from syngas purpose.

As the plasma gasification operating temperatures are in the 1500°C range, this process is very effective to treat any types of waste, while destroying toxic or hazardous materials. The flexibility of the technology is given by the possibility of mixing different types of feedstocks, such as MSW which include metals, glass and electronic, tires, medical waste, etc. The great advantage of plasma gasification is the possibility of using a wide range of fuels in a single facility, from refinery waste, to auto shredder residues, such as fluff or construction debris [39].

Plasma torches have been used to melt or cut metal. When applied to different types of feedstock, they efficiently cause organic and carbonaceous materials to vaporize, whereas the remaining inorganic material is melted and collected after being cooled down as vitrified glass, also known as slag. The melted inorganics pour out at the bottom of the gasifier and once cooled down, convert into a safe and stable material. This represents a major advantage over combustion which generates fly ash.

A typical plasma gasification plant [41] would consist of:
- Plasma furnace, which is usually lined with refractory material to withstand high temperatures
- Secondary combustion chamber for complete soot conversion
- Quenching chamber to avoid dioxin and furans formation
- Particulate matter removal equipment and emissions control systems (hydrogen sulfide adsorber, filters and precipitators, activated carbon, etc.)

Steam can be used as gasifying agent for the plasma technology and induce the water—gas shift toward hydrogen generation. In case air is used, because of the high operating temperatures nitrogen oxides would form. An Air Separation Unit would separate oxygen from regular air and inject it into the gasification reactor [39].

One challenge of plasma gasification is using heat exchangers to recover the heat from the hot gases and generate steam at operating temperatures of 1500°C. High temperatures have strain effects on the heat exchanger materials, as explained in more detail in Chapter 6, Critical Development Needs.

Although there are several pilot plans in operation in Japan and Canada [39], the technology is not yet matured and therefore no company is offering turnkey facilities today.

2.6 HYDROTHERMAL PROCESSES—SUPERCRITICAL WATER GASIFICATION

When dealing with high moisture content feedstock, few treatment options are available. The most developed and wide spread technology is anaerobic digestion, which presents great disadvantages such as long treatment time and wet by-products with low economic value that require further treatment.

An alternative to anaerobic digestion for wet biomass conversion into energy and other valuable by-products is hydrothermal treatment. Hydrothermal refers to an aqueous system at temperatures and pressures near or above the critical point of water.

Depending on the operating conditions, the hydrothermal conversion can be classified as carbonization, oxidation, liquefaction or gasification, which would yield different by-products.

2.6.1 Carbonization

Hydrothermal carbonization is the (pre)treatment of lignocellulosic biomass in hot (180−280°C) [4] water at saturated pressure of 2−10 MPa and residence times varying from minutes to hours. The solid product resulted is hydrophobic, carbon-rich similar to lignite and can be easily pelletized. The aqueous phase contains sugars, acids, and carbon dioxide. The solid is also known as hydrochar, separated from the aqueous phase by a mechanical pressing process and has a higher heating value than that of the feedstock.

2.6.2 Oxidation

Supercritical water oxidation is mainly used to treat wastewater and sludge at temperatures around 650°C and low residence time. The organic pollutants will be degraded during the process and heavy metal particles can be recovered. The process by-products are CO_2, H_2O, and N_2.

2.6.3 Liquefaction

Hydrothermal liquefaction (HTL) converts biomass into liquid fuels in the presence of water or water-containing solvent/cosolvent and a catalyst. The process takes place at temperatures of $200-400°C$ and pressures $5-25$ MPa. The liquid by-product of HTL has high oxygen content; hence, it requires further upgrading treatment [42].

2.6.4 Hydrothermal gasification

Hydrothermal gasification is a good option for treating feedstock with moisture content above 30%.

The process could be operated at subcritical conditions, $225-265°C$ and $2.9-5.6$ MPa or at supercritical conditions. At temperatures below $500°C$ and supercritical conditions, usually a catalyst is required, whereas above $500°C$ no catalysts are used.

Alkali metals, such as NaOH, KOH, Na_2CO_3, K_2CO_3, and $Ca(OH)_2$ were found to be efficient catalysts for supercritical gasification. Research studies have found different types of activated carbon and charcoals, such as coal activated carbon, coconut shell activated carbon, macadamia shell charcoal and spruce wood charcoal to be efficient under supercritical conditions [43].

SWG process relies on the changing properties of water in the supercritical region acting as a solvent and catalyst during gasification and also as a reactant in the hydrolysis reactions. This technology is known to generate high gas yields with low tar and char formation. The feedstock can contain up to 80% water [43] and converted via supercritical water gasification would yield less tars and char at relatively low temperatures, compared to other technologies.

The gasification parameters for this process would have to be above the critical point of water, $374°C$ and 22.1 MPa, where water is in its supercritical state and behaves as a homogeneous fluid phase. The gas-like viscosity of water in supercritical state enhances the mass transfer, whereas the liquid-like density would improve its solvation properties [42].

Supercritical fluids in general have unique properties compared to liquids or gases and have the ability to dissolve material which under regular conditions would not be soluble in liquid water or steam.

Water in supercritical state behaves like an organic, nonpolar solvent and becomes completely miscible with gases and many hydrocarbons [42]. Because of the miscibility, homogeneous reactions of water with organic compounds can occur during supercritical gasification, without the presence of any catalyst required during normal conditions. Steam gasification is a process which usually is operated at temperatures above 1000°C, whereas complete gasification of glucose would be achieved at 650°C, 35.4 MPa under supercritical conditions [44]. The glucose is an intermediate hydrolysis product which is obtained from the reaction of cellulose under hydrothermal conditions.

The hydrothermal processes and especially hydrothermal gasification under supercritical conditions present a few great advantages:

- High energy value of the gas product, due to lower concentrations of CO_2. The high amounts of carbon dioxide released during biomass gasification are mostly dissolved in water under the hydrothermal conditions
- High pressure of the gas is an advantage for subsequent transportation, usage and gas cleaning through steam reforming [44]
- Inorganic compounds such as chloride are found in the aqueous product phase, not causing corrosion problems downstream as dry systems have
- Lower amounts of tars generated

As any other technology, the hydrothermal gasification presents some disadvantages. The major disadvantage is the energy input required to heat up the water in a high moisture content feedstock. In this case the heat provided and the efficiency of the heat exchanger plays an important role in the overall efficiency of the system, because sometimes the energy value within the biomass might not be sufficient to heat up the water contained.

REFERENCES

[1] A. Molino, S. Chianese, D. Musmarra, Biomass gasification technology: the state of the art overview, J. Energy Chem. 25 (1) (2016) 10−25.
[2] M.J. Antal, M. Grønli, The art, science, and technology of charcoal production, Ind. Eng. Chem. Res. 42 (8) (2003) 1619−1640.
[3] J.P. Ciferno, J.J. Marano, Benchmarking Biomass Gasification Technologies for Fuels, Chemicals and Hydrogen Production, U.S. Dep. Energy Natl. Energy June (2002) 58.
[4] A. Pollex, A. Ortwein, M. Kaltschmitt, Thermo-chemical conversion of solid bio-fuels, Biomass Convers. Biorefin. 2 (1) (2012) 21−39.
[5] J. LeBlanc, J.F. Quanci, M.J. Castaldi, Investigating secondary pyrolysis reactions of coal tar via mass spectrometry techniques. p. acs.energyfuels.6b02543. Energy Fuels, 2016.
[6] J. LeBlanc, J. Quanci, M.J. Castaldi, Experimental investigation of reaction confinement effects on coke yield in coal pyrolysis, Energy Fuels 30 (8) (2016) 6249−6256.

[7] C. Marculescu, S. Ciuta, Wine industry waste thermal processing for derived fuel properties improvement, Renew. Energy 57 (2013) 645—652.

[8] W.C. Park, A. Atreya, H.R. Baum, Experimental and theoretical investigation of heat and mass transfer processes during wood pyrolysis, Combust. Flame 157 (3) (2010) 481—494.

[9] I. Turner, P. Rousset, R. Rémond, P. Perré, An experimental and theoretical investigation of the thermal treatment of wood (*Fagus sylvatica* L.) in the range 200—260°C, Int. J. Heat Mass Transf. 53 (4) (2010) 715—725.

[10] C. Gomez, E. Velo, F. Barontini, V. Cozzani, Influence of secondary reactions on the heat of pyrolysis of biomass, Ind. Eng. Chem. Res. 48 (23) (2009) 10222—10233.

[11] I. Milosavljevic, V. Oja, E.M. Suuberg, Thermal effects in cellulose pyrolysis: relationship to char formation processes, Ind. Eng. Chem. Res. 35 (3) (1996) 653—662.

[12] J.E. White, W.J. Catallo, B.L. Legendre, Biomass pyrolysis kinetics: a comparative critical review with relevant agricultural residue case studies, J. Anal. Appl. Pyrolysis 91 (1) (2011) 1—33.

[13] D. Neves, H. Thunman, A. Matos, L. Tarelho, A. Gómez-Barea, Characterization and prediction of biomass pyrolysis products, Prog. Energy Combust. Sci. 37 (5) (2011) 611—630.

[14] J. Ratte, F. Marias, J. Vaxelaire, P. Bernada, Mathematical modelling of slow pyrolysis of a particle of treated wood waste, J. Hazard. Mater. 170 (2—3) (2009) 1023—1040.

[15] S. Singh, C. Wu, P.T. Williams, Pyrolysis of waste materials using TGA-MS and TGA-FTIR as complementary characterisation techniques, J. Anal. Appl. Pyrolysis 94 (2012) 99—107.

[16] C. Casajus, J. Abrego, F. Marias, J. Vaxelaire, J.L. Sánchez, A. Gonzalo, Product distribution and kinetic scheme for the fixed bed thermal decomposition of sewage sludge, Chem. Eng. J. 145 (3) (2009) 412—419.

[17] S. Ciuta, F. Patuzzi, M. Baratieri, M.J. Castaldi, Biomass energy behavior study during pyrolysis process by intraparticle gas sampling, J. Anal. Appl. Pyrolysis 108 (2014) 316—322.

[18] F. Patuzzi, S. Ciuta, M.J. Castaldi, M. Baratieri, Intraparticle gas sampling during wood particle pyrolysis: methodology assessment by means of thermofluidynamic modeling, J. Anal. Appl. Pyrolysis 112 (2015) 1—402.

[19] S.S.E. Ranzi, A. Cuoci, T. Faravelli, A. Frassoldati, G. Migliavacca, S. Pierucci, Chemical kinetics of biomass pyrolysis, Energy Fuels 22 (6) (2008) 4292—4300.

[20] K.S.C. Li, Kinetic analyses of biomass tar pyrolysis using the distributed activation energy model by TG/DTA technique, J. Therm. Anal. Calorim. 98 (1) (2009) 261—266.

[21] C. Li, C.S. Henry, M.D. Jankowski, J.A. Ionita, V. Hatzimanikatis, L.J. Broadbelt, Computational discovery of biochemical routes to specialty chemicals, Chem. Eng. Sci. 59 (22—23) (2004) 5051—5060.

[22] P.R. Patwardhan, D.L. Dalluge, B.H. Shanks, R.C. Brown, Distinguishing primary and secondary reactions of cellulose pyrolysis, Bioresour. Technol. 102 (8) (2011) 5265—5269.

[23] C. Di Blasi, M. Lanzetta, Intrinsic kinetics of isothermal xylan degradation in inert atmosphere, J. Anal. Appl. Pyrolysis 40—41 (1997) 287—303.

[24] J.J.M. Orfão, F.J.A. Antunes, J.L. Figueiredo, Pyrolysis kinetics of lignocellulosic materials—three independent reactions model, Fuel 78 (3) (1999) 349—358.

[25] A. Malik, S.K. Mohapatra, Biomass-based gasifiers for internal combustion (Ic) engines—a review, Sadhana 38 (2013) 461—476.

[26] C. Chen, Y.-Q. Jin, J.-H. Yan, Y. Chi, Simulation of municipal solid waste gasification in two different types of fixed bed reactors, Fuel 103 (8) (2013) 58—63.

[27] C. Lucas, D. Szewczyk, W. Blasiak, S. Mochida, High-temperature air and steam gasification of densified biofuels, Biomass Bioenergy 27 (6) (2004) 563−575.

[28] R.W.R. Zwart, B.J. Vreugdenhil, Tar Formation in Pyrolysis and Gasification, no. June, p. 37, 2009.

[29] S. Kumar, S. Suresh, S. Arisutha, Production of renewable natural gas from waste biomass, J. Inst. Eng. Ser. E 94 (August) (2013) 55−59.

[30] N. Sundac, Catalytic Cracking of Tar from Biomass Gasification, Department of Chemical Engineering, Lund University, Lund, 2007.

[31] N.B. Klinghoffer, M.J. Castaldi, A. Nzihou, Influence of char composition and inorganics on catalytic activity of char from biomass gasification, Fuel 157 (2015) 37−47.

[32] J. Fermoso, B. Arias, C. Pevida, M.G. Plaza, F. Rubiera, J.J. Pis, Kinetic models comparison for steam gasification of different nature fuel chars, J. Therm. Anal. Calorim. 91 (3) (2008) 779−786.

[33] M. Ducousso, E. Weiss-Hortala, A. Nzihou, M.J. Castaldi, Reactivity enhancement of gasification biochars for catalytic applications, Fuel 159 (x) (2015) 491−499.

[34] B. Parra, C.D.E. Sousa, Effect of gasification on the porous characteristics of activated carbons, Carbon N. Y. 33 (6) (1995) 801−807.

[35] C. Pfeifer, S. Koppatz, H. Hofbauer, Steam gasification of various feedstocks at a dual fluidised bed gasifier: Impacts of operation conditions and bed materials, Biomass Convers. Biorefin. 1 (1) (2011) 39−53.

[36] B. Prabowo, K. Umeki, M. Yan, M.R. Nakamura, M.J. Castaldi, K. Yoshikawa, CO_2-steam mixture for direct and indirect gasification of rice straw in a downdraft gasifier: laboratory-scale experiments and performance prediction, Appl. Energy 113 (2014) 670−679.

[37] E.E. Kwon, M.J. Castaldi, Urban energy mining from municipal solid waste (MSW) via the enhanced thermo-chemical process by carbon dioxide (CO_2) as a reaction medium, Bioresour. Technol. 125 (2012) 23−29.

[38] A. Sanna, Advanced biofuels from thermochemical processing of sustainable biomass in Europe, Bioenergy Res. 7 (1) (2014) 36−47.

[39] E. Dodge, Plasma-Gasification of Waste: Clean Production of Renewable Fuels through the Vaporization of Garbage, p. 51, 2008.

[40] E4tech, Review of technologies for gasification of biomass and wastes, Biomass, no. June, p. 125, 2009.

[41] Y. Byun, M. Cho, S. Hwang, J. Chung, Thermal plasma gasification of municipal solid waste (MSW), Gasif. Pract. Appl. (2012) 28. Available from: http://dx.doi. org/10.5772/48537.

[42] O. Yakaboylu, J. Harinck, K.G. Smit, W. De Jong, Supercritical water gasification of biomass: a literature and technology overview, Energies 8 (2) (2015) 859−894.

[43] A. Kruse, Hydrothermal biomass gasification, J. Supercrit. Fluids 47 (3) (2009) 391−399.

[44] L. Guo, C. Changqing, Y. Lu, Supercritical water gasification of biomass and organic wastes, Biomass, no. September 29 (2010) 165−182.

CHAPTER THREE

Laboratory/Demonstration-Scale Developments

Chapter developed in collaboration with:
Marco Baratieri[a] and Francesco Patuzzi[a]

[a]Free University of Bolzano, Bolzano, Italy

Contents

3.1 INTRODUCTION

For a plant to be considered commercial many conditions have to be met, economic benefits of continuously processed large volume and environmental performance. A technology reaches the stage of maturity

37

after long hours of testing on pilot and demonstration-scale units and much skilled labor and capital invested in modifications and optimization. In addition, there must be a commissioning phase that generally lasts about 6–12 months to experience all the challenges associated with long-term operation.

Good collaboration between industries and research organizations through use of laboratory-scale experiments [1] are beneficial for technology advancement and cost reduction. Generally, the industrial organizations focus on operating the unit, whereas the research organizations focus on identifying and understanding the optimal operating conditions. There are several successfully demonstrated large-scale gasification plants, which will be discussed in Chapter 4, Field-Scale Developments. However, pilot and lab-scale gasification and pyrolysis units may help develop the flexibility in design that the market expects for such units. Importantly, one of the advantages posed by gasification and pyrolysis is the ability to produce a syngas that is capable of being upgraded to different products (e.g., liquid fuel or polymers).

The usual path of new technology development starts at bench scale and advances toward pilot sale capacities equal to or greater than 1 t per day [2]. Demonstration-scale facilities prove operation of more than 1000 h and provide data and equipment specification necessary to design the commercial-scale facility [2]. However, very few systems transition from the laboratory or pilot scale to the demonstration stage typically due to finances but often due to technological barriers that are not foreseen during the very initial stages of testing and development.

3.2 BIOMASS FACILITIES

The liquid fuels consumption is projected to increase at an average rate of 1% per year from 2012 to 2040 [3] and biomass to liquid fuels (BtL) is becoming a growing market which is receiving more and more attention from the research and developers communities. The old technologies available, also known as Phase I biorefineries [4] use conventional technologies with very less flexibility of feedstock. These biorefineries use corn, wheat, barley, rye, soybean sugarcane, etc. as feedstock and convert them into chemicals or fuels, such as ethanol. United States are the leading ethanol producer in old generation biorefineries, followed by Brazil, France, China and Canada [5]. Phase II biorefineries can process a variety of lignocellulosic materials and convert them into liquid fuels using more advanced technologies such as Fischer–Tropsch (FT), whereas Phase III biorefineres are more flexible in feedstock process and end product [4]. The CHOREN Sigma Plant in Germany is the only commercial size biomass FT synthesis plant in the world [6]. Other commercial plants that produce liquid fuels via the FT process use municipal solid waste (MSW) or forest and agricultural waste as feedstock. In the United States and other European countries, pilot plants which serve as testing platforms and research facilities offer the possibility to develop and demonstrate technologies. One example is the NREL's Integrated Biorefinery Research Facility that accommodates several pretreatment, enzymatic hydrolysis and fermentation reactors and equipment. Few pilot gasification facilities which convert biomass to liquids or have other uses for the syngas produced are presented in this section.

3.2.1 Gas Technology Institute Gasification Pilot, USA

The Gas Technology Institute (GTI) 24 t/day pilot plant was built in 1974 to conduct near atmospheric pressure gasification testing on coal. The pilot plant underwent upgrades over the years to gasify bituminous coals at 60 psia operating pressure to the development of new 5 t/day unit to gasify coals at pressures up to 510 psia [7]. The GTI gasification technology was licensed to Carbona in 1989, a company which used the same technology to build a coal/biomass gasification pilot plant in Tampere, Finland. This plant processes up to 42 t/day of coal and 60 t/day of biomass at pressures up to 435 psia [7] and combusts the syngas produced to generate district heating for the city of Tampere.

In 2007, the GTI gasification pilot plant was updated to use the oxygen blown Carbona pressurized BFB technology which included gas reforming/cleaning followed by Fisher Tropsch synthesis using cobalt as catalysis [8]. This facility also known as Henry R. Linden Flex-Fuel Test Facility (FFTF) operates at pressures up to 400 psig and is able to gasify coals up to 20 t/day and biomass up to 40 t/day under oxygen agent [8]. The FFTF ran successful tests over the course of years on different types of feedstock from high ash coals from India to wood pellets. The GTI gasification processes in a single stage fluidized bed reactor convert coal (U-GAS) or biomass, pulp mill residues or waste (RENUGAS technology) into syngas.

The GTI gasification technology has been successfully tested on more than 25 types of feedstock [7] listed in Table 3.1 showing the flexibility of

Table 3.1 Feedstock processed at GTI gasification pilot plant [7]

Bituminous coals

Western Kentucky No. 9, Providence
Western Kentucky No. 9 and 11, Camp
Illinois No. 6, Peabody No. 10 and Crown III
Pittsburgh No. 8, Champion and Ireland
Australian, Bayswater No. 2, Sydney Basin
Polish, Silesia
French, Merlebach (run-of-mine)
Utah (run-of-mine)
Colombian
Chinese, Shen Fu
Indian, North Karanpura Coal Field (washed and run-of-mine)

Low rank coals

Montana Rosebud, Colstrip
Wyoming, Big Horn
North Dakota, Freedom
Saskatchewan Lignite, Shand

Coke, char, peat, wastes

Metallurgical coke, US "Bethlehem," Polish, and Chinese
Western Kentucky No. 9 coal char
Illinois No. 6 coal char
Finnish Peat, Viidansuo and Savaloneva
Automobile Shredder Residue

Biomass

Finnish waste wood and paper mill waste
Danish willow
Danish straw
Pelletized alfalfa stems
Pelletized US waste wood
Bagasse

this technology. The process starts with the feedstock being fed into the heated gasifier at temperatures between 840 and 1100°C, depending on the type of fuel. The operating pressure of the gasifier varies from 3 to 30 bar [7]. The gasifier is sometimes operated at temperatures which allow ash agglomeration and selective removal from the bed.

The gasifying agents can be steam, air or oxygen fed into the reactor through a sloping distribution grid at the bottom of the bed, and through an ash discharge port at the center of the distribution grid [7]. The GTI gasification process reaches carbon conversion rates of 95% and cold gas efficiencies of 80% [7].

Ash is removed by gravity at the bottom of the reactor, whereas fly ash is separated by cyclone separators and part of the recovered sly ash is recirculated to the fluidized bed for additional carbon conversion. Fig. 3.1 shows the GTI pilot plant schematic.

This technology reports low levels of tars in the producer gas, below 100 ppmv for C_6 compounds and 5 ppmv for heavier compounds [9]. Sand, limestone, or dolomite are used to maintain the fluidized bed, especially when gasifying biomass.

During 2008–11, the GTI facility was used to demonstrate syngas to biodiesel conversion under a project financed by UPM Finland and E.ON Sweden. During this, 800 h of oxygen gasification testing campaign different catalytic tar reforming catalysts were examined. This phase helped develop the catalytic tar reformer by Haldor Topsoe A/S [10].

Between 2012 and 2014, the Carbona-GTI partnered with Haldor Topsoe, Texas to demonstrate their TIGAS process to convert syngas into alternative fuels. The project had a budget of more than $34 million [11] financed by DOE and converted biomass into fuels by gasification using the Andritz-Carbona process followed by hot gas filtration and tar reforming, acid gas removal by the GTI owned Morphysorb technology and TIGAS gasoline synthesis. This facility was set up to process 21 t/day of wood pellets and generates 23 barrels per day of fuel. The partners of this project have reported [11] tar reforming efficiencies of 85% converted CH_4, 99% conversion of benzene and 99.5% of naphthalene. During the acid removal step, CO_2 values in the syngas were reduced from 40% to 2%. During their testing more than 300,000 lbs of feedstock were gasified and converted into 32,000 lbs of gasoline [10,12]. Carbona Corporation went commercial in 2008 with a 11.5-MWth and 5.5-MW electric project in Skive, Denmark which gasifies wood pellets.

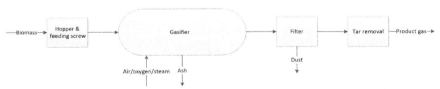

Figure 3.1 GTI Pilot plant schematic. *Adapted from B. Ash, "The GTI Gasification Process," Biomass, pp. 1–9, 2000.*

3.2.2 ICM/Pheonix Bioenergy, USA

The rotary kiln ICM/Phoenix Bioenergy demonstration gasifier was operated at a transfer station in Newton, Kansas from 2009 to 2012 for more than 3200 h, testing various types of biomass, RDF, tire-derived fuel or automobile shredded residue mixed with RDF. The 150-t-per-day facility reported to have tested more than 16 types of feedstock listed in Table 3.2 [13].

The gasification process consists of a horizontal cylinder with an internal auger which slowly rotates [15] allowing feedstock to move through the reactor, whereas air is injected at multiple points. Only small portion of the syngas was used to produce steam, whereas the rest was flared (Fig. 3.2).

Unfortunately, ICM had to take down the demonstration gasifier at the transfer station, upon completion of the project and financing grant, declaring that the facility did not prove to be a viable solution for the county. Some of the problems that ICM mention [16] were related to the availability of feedstock of only 90 t per day, whereas the prototype was designed for 150 t per day, but also insufficient investment from financial partners due to the lower projected returns. ICM announced that through a contract with the City of San Jose, CA they will have the ICM demonstration gasifier at the San José-Santa Clara Regional Wastewater Facility [17]. The facility will process 10 short tons per day of woody

Table 3.2 Feedstock processed at ICM gasification pilot plant [14]

RDF from MSW
RDF blended with tire-derived fuel
Wood chips
Bark
Corn stover and cobs
Rice straw or wheat straw
Oat hulls
Bran
Energy crops
Urban trimmings and wood waste
Construction & demolition waste (C&D)
Bio-refining residue
Sugar cane bagasse
Poultry litter
Biosolids blends
Paper sludge blends

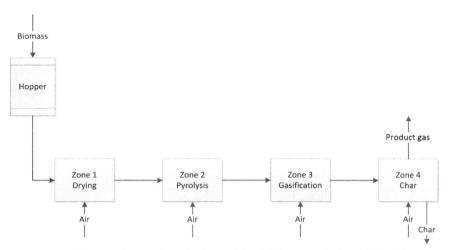

Figure 3.2 ICM gasifier. *Adapted from R.B. Williams and S. Kaffka, "Biomass Gasification—DRAFT," Public Interes. Energy Res. Progr., 2015.*

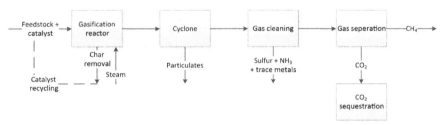

Figure 3.3 Bluegas technology. *Adapted from GreatPoint Energy, "GreatPoint Energy," Int. Adv. Coal Technol. Conf., June, 2010.*

biomass, yard waste or construction and demolition materials mixed with biosolids from the WWT.

3.2.3 BLUEGAS, Great Point, USA

GreatPoint Energy developed the Bluegas technology which converts coals and biomass through catalytic hydromethanation into methane rich syngas. The process flow diagram is shown in Fig. 3.3 [18].

GreatPoint Energy performed multiple tests at the GTI pilot plant in Des Plaines, IL on a large range of feedstock and then constructed and operated a pilot facility in Somerset Massachusetts.

The technology is currently being tested in a pilot plant at the Energy and Environmental Research Center in Grand Forks, ND to demonstrate the technology before designing commercial scale. According to [19], a

$1.25 billion partnership was signed between GreatPoint Energy and China Wanxiang Holding to build a large-scale plant in China. The blue-gas catalytic processes allow for hydromethanation reactions to take place at lower gasification temperatures. Oxygen and steam are directly injected into the fluid bed hydromethanation reactor [20] which drive the catalytic gasification reactions and convert the feedstock into syngas. The methane along with the other gases produced are treated for acid removal and converted into natural gas in a trim methanator using conventional catalytic methanation processes.

3.2.4 VTT Bioruukki, Finland

The VVT Bioruukki Technical Research Centre of Finland has several gasification pilot facilities which are used to test new types of feedstock and to develop new gas cleaning technologies for different applications. One of the pilot plants is a 80-kg/h dual fluidized bed (DFB) steam gasification unit, which is operated at atmospheric pressure and is equipped with hot filtration and gas reforming. Another is a 5-kg/h fluidized bed gasifier, using air, O_2 and steam also equipped with filtration and reforming. VTT started in 2016 the construction of a 80-kg/h pressurized fixed bed pilot plant and a 5-kg/h circulating fluidized bed (CFB) gasification process development unit. The center has a 150-kg/h CFB gasification pilot plant, using steam-O_2 at 7 bar pressures. This unit is presently idle probably whereas the research center is presently focusing on the 2.7-M€ project to convert biomass into transport fuels [21].

Hot gas filtration is applied at temperatures of $750-850°C$ and pressures of 5 bar to syngas to remove particulates, alkali, heavy metals and solid chlorides before downstream reforming. The VTT gas reformer is a staged reformer with a zirconia-based catalyst prereformer before nickel or precious metal-based catalyst stages [22].

3.2.5 Technical University of Vienna, Austria

A fluidized bed gasification test facility is located in Austria, at Technical University of Vienna (TU Vienna). In 2015, the university added a new twin fluidized bed gasification pilot installation, which separates the process into two chambers. The feedstock enters the first high temperature (approx. $850°C$) chamber and is gasified to syngas in the presence of water vapor; then travels to the second chamber where the remaining solids are burned in the presence of oxygen. The temperatures in the gasification chamber reach 850°C and are only 40−80°C lower than the combustion chamber [23]. The heat released from combustion is used to heat up the first chamber by means of hot sand, which circulates between the chambers [24]. The product gas of gasification is rich in hydrogen and has a moderate LHV of 2−14 MJ/N m³, whereas the gas from the second chamber is conventional flue gas.

Superheated steam injected in the first chamber creates a bubbling fluidized bed, whereas in the second chamber the air is injected from the top and the bottom of the combustion chamber. Low concentrations of tar have been reported, ranging from 2 to 6 g/N m³ of gravimetric tar [25] (Fig. 3.4).

The DFB gasification technology has a long history at TU Vienna. Development of the technology started in 1990 and has been since demonstrated at 8 MW$_{th}$ size plants in Güssing and Oberwart, Austria and at a 15-MW$_{th}$ gasifier in Villach, Austria [23]. The pilot plant at TU Vienna is designed to test different types of feedstock, and due to the multiple screw feeders provides the flexibility to gasify a wide range of materials with different particle size and densities [23]. It is known to handle biomass, coal or plastics. The research in the gasification field carried out at the TU of Vienna has significantly advanced the DFB technology making it a more economical and feedstock flexible technology [26]. So far, only feedstocks with high ash content have been shown to create operational problems. Lower tar generation was observed at high operating temperatures and in the presence of steam. Tars which get carried out of the gasification chamber with the remaining solid feedstock are burned in the combustion chamber. The testing carried out at the TU Vienna

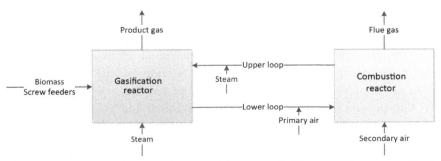

Figure 3.4 TU Vienna pilot plant. *Adapted from V. Wilk, "Extending the range of feedstock of the dual fluidized bed gasification process towards residues and waste," Vienna University of Technology, 2013.*

pilot plant lead to the optimization of all operating parameters, as well as of the optimal moisture of content of the feed material. To obtain a rich hydrogen syngas, different catalytic active beds, such as limestone have been tested [26].

The DFB technology at TU Vienna in partnership with Güssing Renewable Energy GmbH (GRE) advanced this technology even further to successfully process MSW and dried sewage sludge [27]. This technology under the name of Carbon Recyling has at its core the same interconnected gasification and combustion reactors. The product gas runs an Organic Rankine Cycle or thermoelectric generator to produce electricity. After cooling and recovering the residual heat, the gas is cleaned for particulate and tars and can be used to generate synthetic transport fuels or renewable synthetic natural gas (BioSNG) [27].

3.2.6 MILENA Pilot Plant, Netherlands

Energy Research Centre of the Netherlands (ECN) developed and tested the MILENA gasification process in a 30-kW_{th} lab-scale reactor put in operation in 2004 and in the 800-kW_{th} pilot plant put in operation in 2008, both operated for over 5000 hours [28]. The pilot plant converts biomass into a quality syngas with LHV of 16 MJ/N m^3. This CFB gasification technology has been developed over the course of 12 years and consists of three parts: gasifier riser, settling chamber and downcomer [29]. The process is an indirectly heated, allothermal gasification process.

Biomass is fed into the gasifier, whereas superheated steam is injected from below. A hot bed of materials, such as sand or olivine, circulates from the combustion side to the gasification side, heating up the latter section to 850°C. The remaining solid material from the gasification process is carried out to the combustion section together with the bed of material. In the settling chamber gas gets separated from solids entrained with the gas flow, which fall into the down-comer whereas the gas leaves the reactor at the top. In the combustion section, all the solids and tars burn in the presence of air at temperatures of 925°C.

This process is similar to the above mentioned TU Vienna technology and to the Battelle/SilvaGas decommissioned plant described below. The Milena special feature is the settling chamber, which allows for higher residence time and lower tar concentrations. A demonstration plant to produce bio-methane from wood biomass was scheduled to start construction in Alkmaar in 2015. The ECN Milena process, licensed to

Royal Dahlam in combination with the OLGA gas cleaning technology will be used at this 4 MW facility and expected to produce 2.6 million m^3/year of bio–methane.

The OLGA cleaning technology consists of several steps of liquid oil scrubbing, which remove heavy and light tars; wet cleaning, which first removes sulfur and chloride, followed by optional catalytic conversion of syngas into CH_4, CO_2, and water. The final product gas, after removing CO_2 and water has the quality of natural gas and can be injected into the national grid. Before entering the Olga cleaning steps, the gas is cooled down by passing through a cyclone for particulate removal [29].

OLGA gas cleaning was demonstrated at several plants, at ECN air blown gasifiers at a capacity of 2 and 200 m^3/h and on a gasifier at Eneria in France at a capacity of 2000 m^3/h [30]. The Olga gas cleaning process was demonstrated at a gasification plant in Tondela, Portugal which gasifies chicken manure to run a 1–MW_e Caterpillar gas engine. The tar removal efficiencies reported for wood gasification [31] are over 99.9% with phenol and naphthalene concentrations below detection limit. The tar removal efficiency for cleaning the product gas from RDF gasification was reported to be 70.8% [32]. This cleaning technology also removes 30% of benzene and 65% of toluene from producer gas [30]. The plant was retrofitted in 2014 to gasify RDF in CFB reactor and is currently testing the Olga gas cleaning for this feedstock.

Milena technology has demonstrated lower tar concentrations and PM in the product gas, as well as less chance of agglomeration in the combustion chamber. The hot bed materials make a good heat carrier and prevent local hot spots [33].

3.2.7 Verena KIT

The Karlsruhe Institute of Technology (KIT) in Germany is another well-known institute recognized for high-quality research regarding a number of thermal conversion systems. The Verena pilot plant at KIT converts up to 100 kg/h of agricultural biomass and food industry waste at supercritical conditions into a rich hydrogen syngas. The operating conditions are temperatures up to 700°C and pressures of 300 bar. The product gas can be further used in gas turbines, gas engines or fuel cells for power generation. The feedstock running through the system can contain up to 20% dry organic matter and if necessary is mixed with water and grinded [34]. Depending on the level of contamination of the feedstock, additional pretreatment methods might be required. Contamination such as plastics or metals is recommended to be screened out and bulky biomass to be reduced in size. The resulting mixture is compressed to 300 bar using a pump and is heated prior to feed into the reactor using a heat-exchanger followed by a preheater. The physical characteristics of the feedstock mixture need to be optimized for pumping, decreasing the particle size to less than 1 mm [35] and adding enough water.

The reactor is formed from two parts, the tubular reactor which acts like a heat exchanger and a 35-L slim vessel which allows the feedstock residence times of 1 min [36]. At the same time, a stream of fresh water is heated in the heat exchanger and preheater and then introduced in the reactor together with the feedstock mixture at a temperature of approximately 600°C. The product gas that exists the bottom of the reactor is used to heat up the biomass feed by passing through the countercurrent heat exchanger and then cooled down using cooling water. The gas is separated from the water once it cools down and is passed through a separation system followed by CO_2 scrubbing. The process water is sent to a different separator to remove lean gas and eventual solids and then can be reused [36], or stored in tanks (Fig. 3.5).

Obtaining high thermal efficiencies for the supercritical gasification process is always a challenge. According to KIT the thermal efficiency of the process is 80%. The critical aspect is in reusing the product gas as a preheating source of the feedstock via a very efficient heat exchanger.

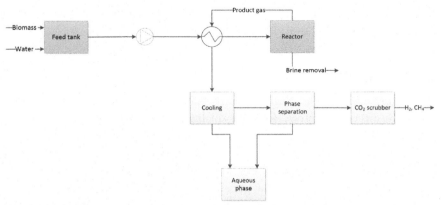

Figure 3.5 Verena KIT gasification process. *Adapted from N. Boukis, U. Galla, H. Müller, and E. Dinjus, "Biomass gasification in supercritical water. Experimental progress achieved with the Verena pilot plant." 15th Eur. Biomass Conf. Exhib. 7–11 May 2007, Berlin, Germany, May, pp. 1013–1016, 2007.*

Alternative solutions for external heating sources could be used to achieve high thermal efficiencies. For example the State Key Laboratory of Multiphase Flow in Power Engineering (SKLMF) of China [37] proposes a solution which couples solar energy used for heating the supercritical water gasification reactor.

Other major challenges of the supercritical water gasification technologies, which have caused the failure of them to become commercial, are related to the clogging of transfer lines. Salt precipitation occurs under supercritical conditions due to the decrease in the salts solubility, leading to clogging problems [37]. To avoid these blockage issues, the Verena pilot plant allows for the separation of brines and solids which fall by gravity at the bottom of the reactor [35]. Table 3.3 shows the difference between the pilot-scale plants described in this section [38].

Table 3.3 Pilot-scale gasification reactors

Technology name	Milena Lab Scale	VTT Pilot Plant	Verena KIT	TU Vienna
Technology type	BFB	CFB	Supercritical Gasification	DFB
Capacity	6 kg/h	80 kg/h	100 kg/h	100 kW
Operating temperature	850−900°C	902−926°C	540°C	850°C
Feedstock type	Beech Wood	Forest wood residue	Corn Silage/ Ethanol	Wood pellets
Bed material	Olivine	Dolomite + Sand	NA	Olivine
Water/steam	Steam: 0.1−2 kg/h	Steam/Fuel Ratio 0.76	Water: 80 kg/h	Steam: 14−16 kg/h
H_2	27.3%	27.3%	77%	40%−43%
CH_4	9.5%	7.9%	15%	8.5%−10%
CO	27.5%	16.5%	1%	26%−28%
CO_2	24.8%	40.1%	2%	19%−21%
Tar	18.3 g/N m^3	6.5 g/N m^3	N/A	2−6 g/N m^3 gravimetric tar 5−15 g/N m^3 GCMS tar

Adapted from B.J.V.C.M. van der Meijden, H.J. Veringa, A. van der Drift, "The 800 kWth allothermal biomass gasifier MILENA," in *16th European Biomass Conference*, 2008, no. June; H. Esa Kurkela, Minna Kurkela, "The Effects of Wood Particle Size and Different Process Variables on the Performance of Steam-Oxygen Blown Circulating Fluidized-Bed Gasifier," *Environ. Prog. Sustain. Energy*, 33, 2, pp. 681−687, 2014; N. Boukis, U. Galla, H. Müller, and E. Dinjus, "Biomass gasification in supercritical water. Experimental progress achieved with the Verena pilot plant.," *15th Eur. Biomass Conf. Exhib. 7−11 May 2007, Berlin, Germany*, May, pp. 1013−1016, 2007; V. Wilk, "Extending the range of feedstock of the dual fluidized bed gasification process towards residues and waste," Vienna University of Technology, 2013.

3.2.8 Other successful biomass pilot facilities

ThermoChem Recovery International (TRI) is an example of successful pilot plant in Durham, North Carolina which has licensed their steam reforming technology to Fulcrum Bioenergy to build a commercial 10 million gallon per year biofuel plant in Nevada [12]. The pilot plant run over 9000 h of steam reforming and 4500 h generating biofuels, testing several types of biomass and MSW.

Pearson Technology, Inc. built a 30-t-per-day biomass to ethanol facility. The gasifier in North Mississippi is converting wood waste in a multistage steam reformer into syngas and into ethanol through an additional FT synthesis stage. The company reports 81% cold gas efficiency and single-pass conversion to ethanol from 15% to 60% [39].

3.2.8.1 Biomass Gasification Research at Iowa State University

The **Iowa State University** combustion and gasification lab is equipped with an atmospheric pressure 7 kW bubbling fluidized bed gasifier and also operates a 900-kW pilot-scale gasification facility approximately six miles away from campus. The **BECON** (Biomass Energy and Conversion facility) converts 5 t of feedstock per day into synthetic natural gas [40]. At present, Iowa State University is working on a project to produce hydrogen rich gas in a ballasted gasifier with steam injection and use of the product gas in fuel cells [40].

The BECON facilities serves as a platform to help researchers and technology developers test different types of feedstock under various conditions. This facility has been operating since 1995, and in 2013 Frontline BioEnergy, LLC partnered with the Iowa State University and the Iowa Energy Center to build the TarFreeGas system [41]. The facility is running to demonstrate the technology of converting biomass into methanol. Other companies using the BECON facility at Iowa State University are: MycoMeal producing ethanol by using fungus to improve conversion efficiency; Innovative Energy Solutions, Inc. is converting plastics, refinery residues or used oils into transportation fuels; Catilin, Inc. is producing biodiesel fuel by using a catalyst [41].

3.2.8.2 GAYA Project—GDF SUEZ

The GAYA Project launched in 2010, coordinated by GDF Suez has the aim of researching and developing synthetic natural gas made from renewable energy sources such as biomass. The 11 industrial and academic partners of this project are currently working on developing a demonstration platform near Lyon, France that will convert biomass into second generation fuels through gasification and methanation.

According to one of the partners Ropotec, Austria, the technologies used at the demonstration plat will be Fast Internally Circulation Fluidized-Bed gasification and synthesis of methane, a process developed and demonstrated by Ropotec in collaboration with Swiss Paul Scherer Institute. According to the project website [42], the commissioning of the demonstration plant should have started in January 2016.

3.3 DISCONTINUED BIOMASS PROJECTS

Several biomass-to-liquid (BtL) facilities have been discontinued in Europe and the United States. The reasons for this are numerous, however they fall into the general categories of insufficient or discontinuation of funding, poor feedstock management, technological issues and undisclosed reasons.

The **Vapo Oy** project planned for a biodiesel plant in Kemi, Finland was discontinued in 2014 because of lack of financial incentives for renewable traffic fuels after the 2020 EU strategy was adopted [43]. Another Finnish BtL 12 MW gasification demonstration plant, **NSE Biofuels** was operated by Neste Oil and Stora Enso. The plant never became commercial because of lack of funding.

The **Carbo-V** biomass gasification technology was bought by Linde Engineering Dresden GmbH from Choren in 2012. The Carbo-V technology from Choren was operated for approximately 2000 h, but could not be put into stable operation [44]. The technology was modified by Linde by reducing the number of equipment units used by 40%. The gasification technology converts woody biomass into biofuels and was licensed to Forest BtL Oy, a Finnish biofuel developer. Forest ByL Oy was building a 480-MW gasification plant in Kemi, Finland to become operational at the end of 2016. The authors have not found to date, any press release on the commissioning of the Kemi gasification plant.

ZeaChem built a demonstration plant in 2012 with a capacity of producing 250,000 gal of biofuel per year in Oregon [45]. The technology developed by the ZeaChem converted biomass into ethanol using a type of bacteria that helps termites digest wood [46]. The conventional processes uses yeast to convert sugars into ethanol, whereas the ZeaChem uses the Moorella thermoacetica bacteria to convert sugars into acetic acid, without any CO_2 release [47]. The acetic acid is then converted into ethyl acetate whereas for the final step to ethanol, the process uses hydrogen produced from the gasification of the biomass remains. In 2013 the company announced that they scaled back operations at the demonstration plant [45].

3.3.1 Ferco SilvaGas

The pilot gasifier at Battelle's Columbus, OH facilities was coupled to a 200-kW Solar gas turbine during testing of different types of biomass and RDF. The plant had a 10-t-per-day capacity and operated over 20,000 h before it was scaled up to the Burlington demonstration plant. The **Ferco** indirectly heated gasification technology was the largest demonstration plant operation of its type in the United States, in Burlington Vermont. The plant was put into operation in 1999 to demonstrate the fuel production through steam gasification process and the gas turbine power generation in a combined cycle. Designed for 200 t of biomass per day, the demonstration plant was initially producing syngas which was cofired in already existing McNeil boilers [48]. The later developments of the project involved additional gas cleanup, gas compression and run of a gas turbine power generator. Because of unknown problems the plant was not operated continuously in the first 18 months after construction which led to problems with feedstock contracts accumulating over this time. Eventually the plant was operated at overcapacity up to 450 t per day and in the year of 2001 a commercial Ferco gasification plant was scheduled to start up [49]. To date there is no information or evidence that the commercial plant was ever commissioned.

3.3.2 Range fuels

Another unsuccessful project is the biomass into ethanol technology developed by **Range Fuels**. The company built a commercial-scale facility in 2007 which gasifies biomass into syngas, followed by conversion into aliphatic alcohols using molybdenum-based catalyst. The Range Fuels plant received $76 million grant from the US DOE, $6 million from the State of Georgia and $158 million in venture capital from several companies [50], but was officially closed in 2012 with a foreclosure sale for reasons unknown to the public. The facility in Soperton, Georgia was planned to be retrofitted by LanzTech, a New Zeeland company and moved to Roselle, Illinois.

3.4 BIOMASS GASIFICATION FOR SMALL-SCALE CHP

Energy generation systems, and among them biomass-to-energy systems, benefit from economies of scale [51], being large-scale systems usually characterized by higher yields and lower cost per unit. The same consideration applies to refineries and biorefineries, as confirmed by most of the examples reported in this chapter. Nonetheless, biomass feedstock usually contain larger amount of water and are characterized by a lower energy density in comparison with the raw material of petrochemical industry. Therefore, the costs associated with transportation makes the concept of small-scale biorefinery attractive [52], overcoming the complexity of the fuel supply logistics related to the intrinsic feedstock characteristics (e.g., the limited period of availability and the scattered geographical distribution over the territory) [53].

In the case of small-scale facilities, a strategy for reducing the investment costs is the partial processing [52], for example producing an intermediate product which can be then delivered, with lower transportation costs, to a centralized plant for the final refinements. Biomass gasification technology can be seen as an application of this concept, since relatively cheap and simple-to-operate air gasifier can be used for obtaining biofuel, the producer gas, which is highly diluted in nitrogen. Even if this aspect makes it not attractive for subsequent refinements, it can be used for CHP production in internal combustion engines (ICEs) by means of some minor and technically feasible modifications (e.g., the increase of the compression ratio in Otto cycle engines allowed by the higher autoignition temperature of producer gas with respect to gasoline [54]).

In the area of R&D projects, numerous prototype/pilot small-scale CHP units based on biomass gasification have been developed [55−59]. Even more interestingly, in the last years, small-scale biomass gasification power plants based on downdraft gasifier and ICE have been successfully commercialized with a global capital cost of about 5000−10,000 €/kW$_{el}$ and electrical and cogeneration efficiencies of about 20% and 80% [60].

In Europe, in countries like Sweden and Finland that are at the forefront of biomass-based CHP due to the strength of the local forest industry, there has been in the last years an increasing interest in small-scale CHP [61]. Something similar happened in South-Tyrol (Italy), where several entrepreneurs invested in the gasification conversion technology for small-scale applications, also driven by an increase on feed-in tariffs for the exploitation of biomass in small production units [62]. The plant owners are both private entities (i.e., local farmers) and companies (i.e., sawmills) that have access to large amounts of low cost local lignocellulosic biomass.

In small-scale applications, the most widespread technology are based on fixed bed, autothermal, air gasifiers. During the last decade, as confirmed by some of the experiences described in this chapter (e.g., the MILENA gasifier developed at ECN and the Viking gasifier developed at DTU), the main driver of innovation for fixed bed reactor has been the multistage approach to gasification [63]. Staged systems are based on the physical separation in different reactors of the subprocesses (drying, pyrolysis, oxidation, reduction) occurring during the thermochemical conversion.

This is also reflected in some of the biomass gasification technologies available in Europe for small-scale CHP, based on a two-stage configuration. On the one side, the separation of the pyrolysis and gasification steps permits a detailed control of the overall process, which results in high-quality producer gas in terms of composition and tar loads. On the other, the requirement for additional equipment and reactors are connected with higher costs with respect to single stage gasifiers. Consequently, there are several example of technologies on which the innovation, rather than focusing on developing multistage gasifiers, is concentrated on optimizing the modularity of gasification systems and the automation of the process. The peculiarities of the most widespread technologies for small-scale CHP production from biomass gasification currently sold in Europe are presented in the following section.

3.4.1 Spanner RE^2 GmbH

Spanner [64] is by far the technology with the highest commercial perpetration. The reactor has the characteristic that the input enters in a pretreatment area where it is usually dried during a gravity- driven downward movement before being uplifted and transported into the entrance of the downdraft gasifier which again is a cocurrent downdraft gasifier. The technology is commonly referred as "Joos gasifier" as a reference to the name of the principle inventor and patent holder Bernhard Joos. The small size and the compact geometry of the gasifier allows the fast start-up of the engine and thus provides flexibility in respect to the hours of operation but also to the biomass type. Nonetheless, the size of the utilized wood chips should be G30–40 (ref. ÖNORM M 7133:1998 02 01), and there is a maximum percentage of 30% fine matter that is recommended in order to sustain undisrupted operation. The biomass input passes through the drying and gasification stages and all gasification products are exiting the reactor in a combined stream. The flow is assisted by means of a pump because the reactor operates under subatmospheric pressures. Tar is removed primarily through condensation and char is separated through bag filter. Subsequently the char passes through a stage of after-burning where it reacts with oxygen under substoichiometric conditions and its mass is significantly reduced. The final solid product of the after-burning is removed with a cyclone. Although Spanner is a well-established company on a commercial level, the produced char is not of typical yield and composition because it is a product of recirculation and secondary reactions. This gasifier uses a turbocharged Otto engine with eight valves which operated in a CHP mode. This technology is manufactured to produce 30 or 45 kWe of nominal electric power, with corresponding thermal outputs of 73 and 108 kWth, respectively.

3.4.2 Kuntschar Energieerzeugung GmbH

Kuntschar gasifiers [65] are also known as hot char bed gasifiers or glowing bed gasifiers. This patented reactor design favors the development of a thick char bed below the reduction/gasification zone by means of a screening plate which makes possible the development of a cavity under the combustion chamber. In smaller scale facilities, the presence of char favors the production of higher quality gas even at lower temperatures. Characteristically, the temperature in the bed of the gasifier rarely exceeds 550°C but the high concentration of carbon monoxide reaches concentrations of 25% which shows that char has a quasicatalytic effect. In principle Kuntschar gasifiers can process biomass wood chips with high moisture content since the drying takes place in a container that is located onsite and is connected to the infeed of the reactor. The reactor is a cocurrent and downdraft gasifier. An interesting and unique characteristic of this facility is that the char, after being filtered out from the gas by a cyclone, is redirected into gasifier. This recirculation increases the carbon conversion efficiency and accelerates the production of high-quality gas products like carbon monoxide. A second notable attribute is that this technology uses a hot filtering system. The dust and char are separated by steel filters and tar is condensed due to the cooling of the producer gas before utilization in the ICE. The engine is a modified diesel engine with added sparkles which has increased compression ratios and thermal efficiency. The typical biomass wood chips that are fed to this reactor are G50 size with low concentration fine matter (maximum 2%) and moderate to low moisture levels (around 13% or less). The installations have slight variations in size and have a range from 100 to 200 kWe and 230–270 kWth.

3.4.3 Burkhardt Energietechnik GmbH

Burkhardt gasifiers [66] are a result of several integrations of patents and novel designs. Although the feedstock—i.e., wood pellets—is fed from the bottom of the gasifier (with a feeding auger) and the output producer gas exits from the top, the distribution zone of the reactor resembles a cocurrent downdraft gasifier. The bottom-fed pellets are dried and pyrolyzed to produce char, bio-oil and pyrolysis gases. The air enters the gasifier in such a way that slightly fluidizes char which mixes and reacts better with the gases. In addition, the design maximizes the retention time of the fuel in the gasifier and provides more time for the gases to react with the char. The reactor is very sensitive to fuels with high ash content since the fuel path in this reactor allows the ashes to melt on the input part of the gasifier and that could block the input stream. Every hour approximately 110 kg of pellets are used and 180 kWe are produced (along with 270 kWth). The gasifier can process efficiently pellets with up to 10% moisture content. This high electrical output is explained by the application of a dual fuel MAN engine which requires an addition four liters of bio-oil per hour. Another feature of this gasification unit is the integration of a wet scrubber for the removal of tar compounds.

3.4.4 URBAS Maschinenfabrik G.m.b.H.

Urbas [67] represents a relatively typical downdraft gasifier an improved Imbert gasifier. Nonetheless, innovations in the distribution of air in the gasifier are a significant optimization step and a crucial technical improvement for the smooth operation of larger downdraft gasifiers. The feeding system uses a sluice system to avoid back-burning and for controlling the air input. The ash is removed from the bottom of the reactor by the cyclic grate movements. The patented system of Urbas includes several innovations in the filtering processes. The particulate matter is removed by a dry scrubber which is a ceramic-coated filter designed for temperatures up to 300°C, packed in a cylinder form. This gasification unit operates with a hot gas cleaning system and the gas cooling takes place in a heat exchanger located directly downstream of the hot filter. During the heat exchanging process the tar is condensed in an electric chiller. The CHP unit is a Mitsubishi synchronous motor which operates at a rate speed of 1500 rpm. The temperature of the gases is 480°C and is reduced to 120°C after the heat exchanging processes. The hot water that is sent to the district heating network has a temperature of 90°C and a return temperature of 70°C. Several PLC controllers are used in this system for safe and continuous operation.

3.4.5 Hans-Werner Gräbner Behälter- und Apparatebau Holzgasanlagen

Gränber gasifiers [68] are relatively small in size since they have a nominal electrical output of 30 kWe and a nominal thermal output of 60 kWth. Although they can't work continuously on a permanent basis these gasifiers have been proved to be very stable. It has been reported that these gasification units can operate for up to 8 h of continuous operation and up to 6000 h per year. The company provides smaller scale gasifiers starting from 10 kWe of nominal electrical output but also a case of a developed 100 kWe gasifier has been reported. The preferred size of wood chips is G40 to G70 and the reactor does not operate optimally with smaller (and relatively typical) wood chips such as G30. In addition, the recommended upper limit for water content is set to be in the range of 15%—20% and wood chips with higher moisture content may disrupt the smooth operation of the reactor. There is no set restriction for the maximum allowed concentration of fine matter in the wood chips but it is recommended for the end-user to be aware that this factor should be taken into consideration. The system does not have an external drier like other systems, therefore the wood chips should be pretreated or dried before entering the reactor. Alternatively, higher quality wood chips should be used at first place. The reactors are autothermal and manually loaded and this is an aspect that emphasizes the active role of the operator in the optimal performance of the gasification unit. The producer gas is cooled down and cleaned through a multistage dry gas filtration processes before used in an ICE.

3.4.6 Syncraft Engineering GmbH

Syncraft gasifier [69] is self-defined as a "floating fixed-bed" reactor which takes advantage of the compaction forces that are developed from the downward gravity-driven pathway of biomass and the countering upward gas flow. But after close examination, it has been proved to be a much more complicated technology. First, it should be stated that this technology operates ideally with dried wood chips, i.e., with moisture content less than 10%, and an average size of G30—G50. An initial technological distinction is represented by the multistage operation mode. The wood chips are pyrolyzed under mild temperatures that start from 200°C in a separated reactor which has a controlled oxygen deprived atmosphere. The maximum attainable temperature in this reactor is 700°C, nonetheless it rarely exceeds 500°C. The scope of this separate stage of (relatively mild) pyrolysis is to convert the wood chips into tar-rich pyrolysis gases and into char which due to the controlled low temperature of conversion tends to have yields of around 80%. The pyrolysis products are fed into the main reactor that was previously defined as a "floating fixed-bed" reactor. The utilization of char instead of wood chips as the main fuel of the reactor ensures that the operating temperature of the combustion zone is high, usually between 1000 and 1300°C, and thus most of the tars are thermally cracked. In the patented "trumpet-shaped" reduction zone the conversion of char produces high-quality gas. This conversion processes are enhanced by the char bed of the reactor which is permeable and loose. Finally, this design allows the separation of impurities and slag, by removing them from the bottom of the reactor and from the opposite direction of the gas exit. The integration of these novel ideas have made possible the creation of these hybrid fixed/floated bed reactors that have the potential to be scaled-up without facing the difficulties of the conventional technologies. The first Syncraft prototypes were operating at a nominal electric output of 250 kWe but now the commercial possibilities include 200 kWe, 300 kWe, and 400 kWe of nominal electric output. Their electric efficiency can reach up to 30% which is a very high value compared to competing small-scale gasification technologies.

3.4.7 Stadtwerke Rosenheim GmbH

The Stadtwerke Rosenheim technology [70] is another example of a more sophisticated small-scale system (but with scaling-up potential) that is entering the commercial stage. The feeding system is similar to the Joos gasification system where the infeed enters a drying vessel and follows a downward and gravity-driven route before being uplifted to the entry point of the gasifier. From that point several innovations are integrated for the optimization of the performance. The feeding auger that uplifts the dried material into the reactor is surrounded by pipelines which contain the hot gas that exits from the gasifier downstream of the process. Thus, the rotating feeding auger is converted in a pyrolysis reactor which is driven by heat exchange with the hot gas. As a result, the temperature in the feeding auger starts from 200°C and exceeds 700°C in the latter part of the auger. The temperature of the gas drops from 900°C—which is the exit temperature from the gasifier—to 400°C. The residence time of the biomass in the pyrolyzing auger is set to be 20 minutes to ensure the complete conversion of the input into pyrolysis products. The multistage approach of the design allows the development of pyrolysis products, i.e., primarily char, and as a result the temperature in the oxidation zone of the main reactor exceeds 1000°C. This high temperature conditions increase the overall efficiency of the gasifier but are also the cause of ash melting and potential slag formation. In conventional designs these effects would be countered by controlling the input air and thus slightly lowering the temperature of operation. In this case these effects are countered by inherent character-istics of the design of the gasifier, which allows the downward removal (with gravity) of any impurities. The gasifying agent, i.e., air, is injected at the center of the oxidation zone and is uniformly distributed. The producer gas exits at the top of the gasifier and because of the reactor design, a high temperature of 900−950°C is able to be sustained in the gasification zone. An additional innovation that is proposed by Stadtwerke Rosenheim is the potential utilization of multiple feeding augers/pyrolyzers in order to distribute the input of the pyrolysis evenly in the oxidation zone. A typical installation has a nominal electrical output of 50 kWe and thermal electrical output of 95 kWe. A larger gasifier of 200 kWe and 300 kWth is under development

3.4.8 Pyrox GmbH

Pyrox gasifiers [71] are manufactured in various sizes ranging from 300 to 995 kWe nominal electrical output. As an example, the system of nominal electrical output of 850 kWe has thermal output of 1060 kWth. The reactor is following the standardized downdraft fixed bed approach but is utilizing some very interesting innovations in order to improve the performance of the reactor. The gasifier is cylindrical and the size of designed operation is such that does not allow a narrow throat that would disrupt the smooth operation of the gasifier. The innovative aspect of this gasifier is the internal aeration system which is located at the center of the reactor. At first glance, this would resemble to a classical Buck Rogers system but this centralized system is also used to insert the pyrolysis gases directly in the center of the oxidation zone, along with the staged and controlled aeration. This is a direct way to separate the pyrolysis products "in situ" without the addition of preliminary processing stages. Through the pyrolysis gases recirculation the operation of the gasifier is optimized because of two main reasons. First, the combustion is assisted by the addition of the pyrolysis gases and the temperature is increased in the oxidation zone. It has to be denoted that high temperatures in the oxidation zone is not typical for reactors with staged aeration like this one. Another result of this application is that the increased temperature in the oxidation zone cracks the heavier tars. This is an important improvement because bigger downdraft gasifiers usually generate producer gases with high tar concentrations. Therefore, this approach has made possible a successful scaling of the downdraft designs up to the size of 1 MWe.

3.4.9 Xylogas Energieanlagenforschung GmbH

Xylogas gasifiers [72] are consist of different integrated modular parts. This technology utilizes a self-developed and patented dryer that is named "walking floor" where the moisture content of the wood chips is controlled between 12% and 18%. Part of the necessary thermal energy for drying is produced in the CHP engine and the rest is provided from the captured heat during gas cooling. The feeding of the reactor is done from the top and the reactor is a downdraft fixed bed gasifier. The very high temperature of the oxidation zone produces a high-quality gas with high purity. This temperature is achieved by use of the additional recovered heat which is recirculated around the pyrolysis zone. The air input is carefully controlled and as a result the excess oxygen in the final product is negligible. The solid byproducts are removed from the bottom of the reactor. The producer gas is purified in a dry cleaning systems (ESP plus a safety filter) and therefore there is no need for wastewater treatment. The reactor uses wood chips of G50—G100 size. The cleaned gas is combusted in a 12-cylinder MAN engine for CHP production. Xylogas systems are manufactured in four different sizes: 220, 440, 660, and 900 kWe, respectively, while the nominal electrical efficiencies are above 25% for all the cases. As mentioned, Xylogas relies on the scaling up of its modular units, therefore the 440 kWe and the 660-kWe options are achieved by combining 220 kWe units in parallel.

3.5 CONCLUSIONS

The gasification processes, such as the above mentioned Milena and TU Vienna, operated at temperatures lower than 1000°C are known to yield higher amounts of tars. Instead of using catalysts, commonly utilized in other processes for tar cracking, the second combustion stage of the gasification technology serves very well the purpose of burning tars and remaining solids. The Milena process coupled to the Olga tar removal technology, which recirculates the tars back to the Milena combustion chamber, increases the capture efficiency of the overall process.

The gasification processes comes with great challenges no matter the scale. Pilot-scale facilities encounter various problems which usually take a lot of trial and error, time and finances to solve. Using data generated in pilot-scale facilities to model and simulate further developments is a well-known path.

REFERENCES

[1] A. Bridgwater, The Future for Biomass Pyrolysis and Gasification: Status, Opportunities and Policies for Europe, Aston University (November) (2002) 1—30.
[2] DOE, Integrated Biorefineries: Reducing Investment Risk in Novel Technology, U.S. Dep. Energy (April) (2014) 4.
[3] U.S. Energy Information Administration, International Energy Outlook 2016, vol. 0484(2016), no. May 2016. 2016.

[4] A. Sanna, Advanced biofuels from thermochemical processing of sustainable biomass in Europe, Bioenergy Res. 7 (1) (2014) 36—47.

[5] V. Balan, Current challenges in commercially producing biofuels from lignocellulosic biomass, ISRN Biotechnol. 2014 (i) (2014) 1—31.

[6] S.S. Ail, S. Dasappa, Biomass to liquid transportation fuel via Fischer Tropsch synthesis—Technology review and current scenario, Renew. Sustain. Energy Rev. 58 (2016) 267—286.

[7] B. Ash, The GTI gasification process, Biomass (2000) 1—9.

[8] P. Kukkonen, From Biomass to Biobusiness Andritz and Biomass, pp. 1—2, 2008.

[9] A. Horvath, K. Salo, J. Patel, Synthesis Gas Production from Biomass ANDRITZ Company Profile Business Groups, October, 2011.

[10] A.I. Horvath, Synthesis gas generation for transportation fuel production synthesis gas generation for transportation fuels content, Gasif. Technol. Conf 2014 (2014).

[11] R. Knight, Green gasoline from wood using carbona gasification and Topsoe TIGAS processes, DOE Bioenergy Technol. Off. 2015 Proj. Peer Rev. (2015).

[12] K. Whitty, E. Shanin, S. Owen, Biomass Gasification in the United States, 2015.

[13] K. Colwich, ICM, Inc. Announces Contract of its Advanced Gasification Technology with JUM Global for the City of San José, 2015. [Online]. Available: http://www.icminc.com/icm-media/whats-new-at-icm/23-press-releases/224-icm-announces-contract-of-its-advanced-gasification-technology-with-jum-global-for-the-city-of-san-jose.html. [Accessed: 01-Mar-2017].

[14] ICM, Advanced Gasification Technology, 2013.

[15] R.B. Williams, S. Kaffka, Biomass Gasification—DRAFT, Public Interes. Energy Res. Progr (2015).

[16] A. Bergner, ICM Plans to Continue Work on Gasifier Technology, 2013. [Online]. Available: http://www.icminc.com/icm-media/whats-new-at-icm/24-icm-in-the-news/173-icm-plans-to-continue-work-on-gasifier-technology.html.

[17] B. Bennett, Commercial-Scale Demonstration of ICM's Gasification Technology, 2012 Gasif. Technol., 2012.

[18] GreatPoint Energy, GreatPoint Energy, Int. Adv. Coal Technol. Conf. (June) (2010).

[19] NETL, "Great Point Energy," Gasifipedia, 2017. [Online]. Available: https://www.netl.doe.gov/research/coal/energy-systems/gasification/gasifipedia/gpe.

[20] P. Raman, Innovative Catalytic Gasification Technology to Maximize the Value of Wyoming â€™ s Coal Resources, Gt. Energy's V3.0 bluegas™ Hydromethanation Res. Proj.

[21] VTT, BTL2030 Project, 2016.

[22] P. Simell, I. Hannula, S. Tuomi, M. Nieminen, E. Kurkela, I. Hiltunen, et al., Clean syngas from biomass—process development and concept assessment, Biomass Convers. Biorefinery 4 (4) (2014) 357—370.

[23] V. Wilk, Extending the Range of Feedstock of the Dual Fluidized Bed Gasification Process Towards Residues and Waste, Vienna University of Technology (2013).

[24] B. Messenger, Vienna University Opens Twin Fluidised Bed Waste Gasification Test Plant, Waste Management World, 2015. [Online]. Available: https://waste-management-world.com/a/vienna-university-opens-twin-fluidised-bed-waste-gasification-test-plant. [Accessed: 01-Mar-2017].

[25] J.H.R. Rauch, Country Report Austria, 2015.

[26] C. Pfeifer, S. Koppatz, H. Hofbauer, Steam gasification of various feedstocks at a dual fluidised bed gasifier: impacts of operation conditions and bed materials, Biomass Convers. Biorefinery 1 (1) (2011) 39—53.

[27] Gussing Renewable Energy, Gre multifuel gasification, 2014.

[28] ECN, MILENA gasification process, 2008.

[29] B.J.V.C.M. van der Meijden, H.J. Veringa, A. van der Drift, The 800 kWth allothermal biomass gasifier MILENA, in 16th European Biomass Conference, 2008, no. June.

[30] ECN, Performance of OLGA tar Removal System, 2009.

[31] R. Dahlman, Royal Dahlman Demonstration Plant, 2014. [Online]. Available: http://www.royaldahlman.com/renewable/news/royal-dahlman-demonstration-plant-in-portugal/. [Accessed: 01-Mar-2017].

[32] J. Könemann, MILENA-OLGA Integrated Gasification and Gas Cleaning Technology, in The International Conference on Thermochemical Conversion Science, 2015.

[33] ECN, MILENA Biomass Gasification Process, 2011. [Online]. Available: http://www.milenatechnology.com/home/. [Accessed: 01-Jan-2017].

[34] KIT, 2004. Hydrogen and Methane Production from Wet Biomass. [Online]. Available: http://www.ikft.kit.edu/`downloads/boukis-flyer-verena.pdf. [Accessed: 20-Jul-2003].

[35] N. Boukis, U. Galla, H. Müller, E. Dinjus, Biomass gasification in supercritical water. Experimental progress achieved with the Verena pilot plant, 15th Eur. Biomass Conf. Exhib. 7−11 May 2007, Berlin, Germany (May) (2007) 1013−1016.

[36] A. Boukis Nikolaos, Volker Diem, Eckhard Dinjus, Ulrich Galla, Kruse, Biomass gasification in supercritical water, 12th European Conference on Biomass for Energy, Industry and Climate Protection (June) (2002) 1339−1341.

[37] O. Yakaboylu, J. Harinck, K.G. Smit, W. De Jong, Supercritical water gasification of biomass: A literature and technology overview, Energies 8 (2) (2015) 859−894.

[38] I.H. Esa Kurkela, Minna Kurkela, The effects of wood particle size and different process variables on the performance of steam-oxygen blown circulating fluidized-bed gasifier, Environ. Prog. Sustain. Energy 33 (2) (2014) 681−687.

[39] T. Miles, Pearson Technologies, BioEnergy List, 2006. [Online]. Available: http://gasifiers.bioenergylists.org/pearsonref. [Accessed: 20-Jul-2002].

[40] N.C. Alex Bumgardner, B. Staehling, Biomass Gasification Research at Iowa State University, Sci. Eng. Committee, Iowa State Univ (2006).

[41] Iowa Energy Center, "Becon Facility," no. 515, p. 2, 2014.

[42] "Project Gaya." [Online]. Available: http://www.projetgaya.com/en/.

[43] ETIP, European Technology and Innovation Platform. [Online]. Available: http://www.etipbioenergy.eu/. [Accessed: 20-Jun-2017].

[44] H. Kittelmann, Carbo-V® Biomass Gasification Technology: Status after Application of Sound Engineering Practices, IAE Work. (November) (2014).

[45] C. Proctor, ZeaChem scales back at Oregon biofuel plant, Denver Business Journal (2013) [Online]. Available: European Biofuels Technology. [Accessed: 01-Aug-2016].

[46] ZeaChem, ZeaChem Technology Institute. [Online]. Available: http://www.zeachem.com/technology-institute

[47] K. Bullis, 2008. Creating Ethanol from Wood More Efficiently, MIT Technology Review, 2008. [Online]. Available: https://www.technologyreview.com/s/409490/creating-ethanol-from-wood-more-efficiently/. [Accessed: 01-Aug-2017].

[48] M.a Paisley, R.P. Overend, Verification of the Performance of Future Energy Resources 'SilvaGas® Biomass Gasifier—Operating Experience in the Vermont Gasifier, Pittsburgh Coal Conf 3 (2002).

[49] DOE, DOE Report, pp. 163−190, 1999.

[50] Wikipedia, Range Fuels, 2017. [Online]. Available: https://en.wikipedia.org/wiki/Range_Fuels. [Accessed: 01-Oct-2016].

[51] V. Dornburg, A.P.C. Faaij, Efficiency and economy of wood-fired biomass energy systems in relation to scale regarding heat and power generation using combustion and gasification technologies, Biomass Bioenergy 21 (2) (2001) 91−108.

[52] L. Axelsson, M. Franzén, M. Ostwald, G. Berndes, G. Lakshmi, N.H. Ravindranath, Perspective: Jatropha cultivation in southern India: assessing farmers' experiences, Biofuels, Bioprod. Biorefining 6 (3) (2012) 246−256.

[53] A.C. Caputo, M. Palumbo, P.M. Pelagagge, F. Scacchia, Economics of biomass energy utilization in combustion and gasification plants: effects of logistic variables, Biomass Bioenergy 28 (1) (2005) 35−51.
[54] G. Sridhar, Biomass derived producer gas as a reciprocating engine fuel—an experimental analysis, Biomass Bioenergy 21 (1) (2001) 61−72.
[55] D. Rovas, A. Zabaniotou, Exergy analysis of a small gasification-ICE integrated system for CHP production fueled with Mediterranean agro-food processing wastes: the SMARt-CHP, Renew. Energy 83 (2015) 510−517.
[56] U. Lee, E. Balu, J.N. Chung, An experimental evaluation of an integrated biomass gasification and power generation system for distributed power applications, Appl. Energy 101 (2013) 699−708.
[57] Z. Zhou, X. Yin, J. Xu, L. Ma, The development situation of biomass gasification power generation in China, Energy Policy 51 (2012) 52−57.
[58] T.J.B. Warren, R. Poulter, R.I. Parfitt, Converting biomass to electricity on a farm-sized scale using downdraft gasification and a spark-ignition engine, Bioresour. Technol. 52 (1) (1995) 95−98.
[59] P. Hasler, T. Nussbaumer, Gas cleaning for IC engine applications from fixed bed biomass gasification, Biomass Bioenergy 16 (6) (1999) 385−395.
[60] E. Bocci, M. Sisinni, M. Moneti, L. Vecchione, A. Di Carlo, M. Villarini, State of art of small scale biomass gasification power systems: a review of the different typologies, Energy Procedia 45 (2014) 247−256.
[61] M. Salomón, T. Savola, A. Martin, C.J. Fogelholm, T. Fransson, Small-scale biomass CHP plants in Sweden and Finland, Renew. Sustain. Energy Rev. 15 (9) (2011) 4451−4465.
[62] F. Patuzzi, D. Prando, S. Vakalis, A.M. Rizzo, D. Chiaramonti, W. Tirler, et al., Small-scale biomass gasification CHP systems: comparative performance assessment and monitoring experiences in South Tyrol (Italy), Energy 112 (2016) 285−293.
[63] S. Vakalis, M. Baratieri, State-of-the-art of small scale biomass gasifiers in the region of South Tyrol, Waste Biomass Valorization 6 (5) (2015) 817−829.
[64] B. Joos, Device for creating a flammable gas mixture, Patent Number: EP 2522707 A2, Nov-2012.
[65] W. Kuntschar, Co-current gasifier with Hot char Bed, Patent Number: EP 1 616 932 A1, May-2005.
[66] K. Weichselbaum, Method and device for thermochemically gasifying solid fuels, Patent Number: WO 2010046222 A2, Apr-2010.
[67] Peter Urbas, P. Ebenberger, W. Felsberger, Holzvergasungsanlage, Patent Number: WO 2008089503 A1, Jan-2008.
[68] H.S. Gräbner, H.J. Gräbner, Anhang A: Kurzbeschreibung Anlage Gräbner, in Schwachstellenanalyse an BHKW-Vergaseranlagen, 2009.
[69] M.B. Huber, Gasifier (Syncraft Engineering Gmbh), Patent Number: US 20100095592 A1, Mar-2008.
[70] K. Artmann, R. Egeler, G. Kolbeck, W.S. Christian Schmidt, R. Waller, Biomass gasifier (Stadtwerke Rosenheim Gmbh & Co), Patent Number: EP 2641958 A1, Mar-2013.
[71] B. Dietz, Method and shaft gasifier for generating fuel gas from a solid fuel (Pyrox Gmbh), Patent Number: WO 2011095347 A2, Jan-2011.
[72] "Xylogas® Woodgas - power plants." [Online]. Available: www.xylogas.com. [Accessed: 28-Mar-2017].

CHAPTER FOUR

Field Scale Developments

Contents

4.1 INTRODUCTION

Commercialization of waste gasification and pyrolysis technologies is again experiencing a renewed development in the waste management hierarchy. Currently, Japan is the global leader in waste gasification with a relatively long track record of about 15 years for commercial systems [1]. Waste gasification is prevalent in Japan primarily because this island nation has limited land space for landfilling; therefore, thermal waste conversion technologies are essential to the country's waste management framework.

Gasification of Waste Materials.
DOI: http://dx.doi.org/10.1016/B978-0-12-812716-2.00004-2

In addition, Japan's regulations require that waste produced in a prefecture must be treated within that same prefecture. This results in small to moderate size systems (i.e., 10−500 tons per day (tpd)) that are well suited for gasification technologies. Also, since inertash is the desired residual, systems that can vitrify ash (such as gasification processes) are most prevalent. There is a growing interest in Europe and in the United States (U.S.) in waste gasification because of the advantages that syngas production from a gasification process can offer. Specifically, syngas is a versatile fuel product that can be converted to electricity or hydrogen for numerous applications such as liquid fuels and chemicals [2].

Gasification and pyrolysis processes can recover energy and materials from a variety of wastes. Gasification technologies process waste feedstocks such as biomass, municipal solid waste (MSW), sludge and wet solid waste, hazardous waste, and medical waste, and vary based on the gasifier type and the final product yield. Pyrolysis processes can also handle MSW and organic materials; however, most commercial processes typically target nonrecycled plastics (NRP) as their feedstock and convert the NRP to heating and transport oils that can be sold back into the general commercial or industrial market.

Although a primary appeal of waste gasification and pyrolysis is the fuel product flexibility, all established field scale operations currently burn the syngas to produce electricity. The Enerkem facility in Edmonton, Canada is the first commercial waste gasification facility to produce biofuel from refuse derived fuel (RDF) and a technical case study of its pilot plant operations is provided in this chapter. The primary technical challenges in the scale-up of waste gasification and pyrolysis technologies are feedstock handling and syngas clean-up. These technical challenges combined with the high capital costs of waste gasification and pyrolysis facilities are the hurdles that this industry needs to overcome in order to make commercialization more viable. Nonetheless increasing landfilling costs in the U.S. and landfill bans in Europe and Japan open up opportunities for these technologies to step in and diversify the end products from waste conversion. Optimized front-end waste collection, to produce more homogeneous streams, and modular applications will be key in the growth of these technologies as they will enhance their performance and reliability which in turn will make them more cost competitive.

The following sections of this chapter provide an overview of field scale developments in gasification and pyrolysis for biomass, MSW, NRP, and wet solid waste. Technical case studies of the processes of Enerkem,

Golden Renewable Energy (GRE), and Sustainable Waste Power Systems (SWPS) are also included in the chapter. These processes represent a cross-section of the waste streams that are at the forefront of active development and research.

4.2 BIOMASS AND MUNICIPAL SOLID WASTE FACILITIES

4.2.1 Overview

Field scale commercial gasification operations began in the 1920s using coal as a feedstock. The most well-known coal gasification processes are those of Sasol and U-gas [3]. In contrast, gasification using MSW has been tried numerous times but still remains a developing industry. The challenge of waste gasification compared with traditional coal gasification is the variability of the feedstock which in turn makes it more difficult from an operations standpoint to yield a consistent final product in terms of quality and yield. Nonetheless, there are numerous demonstrational and pilot scale waste gasification facilities worldwide designed to process MSW and convert it to electricity or liquid fuels and chemicals. One of the primary challenges in scaling up waste gasification processes to field scale operation is the economics. Japan is currently the leader in commercial field scale waste gasification because the country does not have the space to landfill its waste; therefore, its geographical limitations combined with its waste management infrastructure and government support make waste gasification economically feasible. As mentioned previously, there is a growing interest in Europe and the U.S. in MSW gasification because of the potential to be more efficient and produce syngas that can be further upgraded to specific products. The following includes a detailed summary of the primary leaders in field scale MSW gasification and an overview of companies worldwide that are pursuing field scale gasification of MSW and biomass.

4.2.2 Staged gasification

Staged gasification refers to gasification technologies that immediately combust the syngas in a direct-connect downstream boiler to produce electricity. This is the most common form of field scale gasification operations because it does not require extensive syngas clean-up of tars and acid gases. The combustion results in a flue gas that is very similar, if not the same, as the gas produced from direct combustion of the feed-stock and can be cleaned using existing air pollution control (APC) systems. The advantage of staged gasification compared to direct waste combustion is that higher efficiencies may be achieved due to homogenous gas—gas reactions and reduced total flue gas volumetric flowrate [4].

4.2.2.1 Energos

Energos, which was formerly a Norwegian supplier and now operates under the British company Ener-G, uses grate gasification of MSW for combined heat and power (CHP) systems to produce grid electricity and district heating from refuse derived fuel (RDF) [4]. MSW feedstock is shredded to approximately <150 mm particle size prior to feeding into the gasifier [3]. Gasification occurs in air at 900°C and combustion of the syngas occurs at 1000°C. Approximately 30% of the flue gas generated is recirculated to heat the Energos system [4]. Energos has built 7 facilities across Europe, specifically in Norway, Germany, and the United Kingdom, with the first facility commissioned in 1997 [5]. However, Energos has recently filed for bankruptcy due to cash flow problems related to its strategy of waste procurement and renewable energy credits. Table 4.1 lists the Energos operational facilities that process MSW and their annual capacities.

4.2.3 Slagging gasification

Slagging gasification refers to gasification technologies that vitrify the ash residual of the process. Japan predominantly operates slagging gasifiers because vitrification reduces the volume of residual that needs to be disposed. Vitrification also renders the ash inert; therefore, it can be used in materials applications such as in the production of concrete.

4.2.3.1 Nippon Steel

Nippon Steel is the largest supplier of gasification plants in Japan and its facilities primarily employ fixed bed updraft gasification. The gasification occurs in oxygen-enriched air (approximately 36% O_2) at 1800°C, and coke is added to achieve high temperatures to vitrify the residue. The syngas is combusted in excess air in a combustion chamber at approximately 1100°C and the final products of the process are electricity and slag. Nippon Steel currently has 33 facilities in operation with an additional 5 facilities under construction. The facilities are located in Japan

Table 4.1 Energos MSW gasification facilities

Plant name	Location	Commissioned in	Design capacity (ton/year)	Output
Averoy	Norway	2000	30,000	CHP 69 GWh (thermal/year)
Hurum	Norway	2001	39,000	105 GWh (thermal/year)
Minden	Germany	2001	39,000	105 GWh (thermal/year)
Forus	Norway	2002	39,000	CHP 105 GWh (thermal/year)
Sarpsborg 1	Norway	2002	39,000	CHP 105 GWh (thermal/year)
Isle of Wight	United Kingdom	2009	30,000	1.8 MW (electrical/year)
Sarpsborg 2	Norway	2010	78,000	256 GWh (thermal/year)

and Korea and range in capacity from approximately 30,000 to 200,000 tonnes per annum (tpa) [1]. Table 4.2 lists the Nippon Steel operational facilities that process MSW and their annual capacities.

4.2.3.2 Ebara

Ebara, a Japanese company, employs a fluidized-bed gasification technology called the pressurized twin internally circulating fluidized-bed gasification system, also commonly known as Twin Rec [6]. Twin Rec is a staged gasification process for RDF that produces electricity from syngas combustion and slag. The fluidized-bed gasifier operates in air at 600°C and the resultant syngas and high carbon residual are then sent to a cyclonic combustion chamber operated at 1400°C. The high combustion temperature vitrifies the ash making it inert [4]. There are currently 12 Ebara facilities operating to treat MSW and industrial waste with 3 of them in Korea and the rest located in Japan [3]. Table 4.3 lists the Ebara operational facilities that process MSW and their annual capacities.

4.2.3.3 Thermoselect

The Thermoselect technology is licensed to JFE Group of Japan and Interstate Waste Technologies of the U.S. The technology consists of a pyrolysis stage followed by high-temperature melting and gasification that

Table 4.2 Nippon Steel MSW gasification facilities

Plant location	Year	Capacity (tpd per line)	Number of lines	Annual capacity (tpa)
Kamaishi, Iwate	1979	50	2	30,000
Ibaraki, Osaka [1]	1980	150	3	135,000
Ibaraki, Osaka [2]	1996	150	2	90,000
Iryu, Hyogo	1997	60	2	36,000
EIFU, Kagawa	1997	65	2	39,000
Iizuka, Fukuoka	1998	90	2	54,000
Ibaraki, Osaka	1999	150	1	45,000
Itoshima, Fukuoka	2000	100	2	60,000
Kameyama, Mie	2000	40	2	24,000
Akita	2002	200	2	120,000
Maki, Niigata	2002	60	2	36,000
Kazusa, Chiba	2002	100	2	60,000
EIFU, Kagawa	2002	65	1	19,500
Takizawa, Iwate	2002	50	2	30,000
Narashino, Chiba	2002	67	3	60,300
Kochi West	2002	70	2	42,000
Tajimi City, Gifu	2003	85	2	51,000
Toyokawa, Aichi	2003	65	2	39,000
Oita City	2003	129	3	116,100
Munakata, Fukuoka	2003	80	2	48,000
Seinoh, Gifu	2004	90	1	27,000
Shimada City, Shizuoka	2006	74	2	44,400
Kazusa, Chiba	2006	125	2	75,000
Kita-kyushu, Fukuoka	2007	240	3	216,000
Yangsan City, Korea	2007	100	2	60,000
Fukuroi, Shizuoka	2008	66	2	39,600
Narumia, Aichi	2009	265	2	159,000
Goyang City, Korea	2009	150	2	90,000
Shizuoka	2010	250	2	150,000
Himeji, Hyogo	2010	134	3	120,600
Matsue, Shimane	2010	85	3	76,500
Okazaki, Aichi	2011	190	2	114,000
ISCEU, Iwate	2011	73.5	2	44,100

yields syngas and slag. Pyrolysis is operated at 600°C and gasification is operated at 2000°C in oxygen to yield an inert slag. The thermoselect technology processes as-received MSW and yields a syngas that can be used in a gas turbine or converted to liquid fuels and chemicals [3]. It claimed that the gas cooling stage of this process suppresses the generation of dioxins and furans [7]. There are currently 6 operational Thermoselect

Table 4.3 Ebara MSW gasification facilities

Plant location	Year	Capacity (ton/day per line)	Number of lines	Annual capacity (tpa)
Sakata	2002	98	2	58,800
Kawaguchi	2002	140	3	126,000
Ube	2002	66	3	59,400
Chuno	2003	56	3	50,400
Minami–Shinshu	2003	46.5	2	27,900
Nagareyama	2004	69	3	62,100
Shiga	2007	60	3	54,000
Taegu City, Korea	2008	70	1	21,000
Eunpyeong, Korea	2009	48	1	14,400
Hwasung City, Korea	2010	150	2	90,000
Okinawa	2010	103	3	92,700
Gimpo, Korea	2012	42	2	25,200

Table 4.4 Thermoselect/JFE MSW gasification facilities

Plant location	Began operation	Capacity (ton/day)
Kurashiki, Japan	2005	550
Yorii, Japan	2006	450
Ishhaya, Japan	2005	300
Mutsu, Japan	2003	140
Tokushima, Japan	2005	120

facilities in total and they are all located in Japan. Table 4.4 lists the Thermoselect/JFE operational facilities that process MSW and their daily capacities.

4.2.4 Plasma gasification

Plasma gasification is advantageous in that its high-temperature operations have the potential to result in high efficiencies with reduced emissions compared to other gasification technologies. However, plasma gasification is very energy intensive and care must be taken to fully integrate and harness the high-grade energy produced. Plasma systems were originally developed for the destruction of medical and hazardous wastes where energy recovery costs were not an issue. However, recent efforts have been focused on adapting plasma technology for MSW and sludge. Plasma gasification units must preprocess the waste material either via size reduction or with the addition of auxiliary fuel.

4.2.4.1 AlterNRG

AlterNRG is a Canadian company that uses the Westinghouse plasma-assisted gasification technology to convert MSW into electricity and slag. The waste does not pass through the plasma torch but is instead heated indirectly and gasified using oxygen and steam at a bulk bed temperature of approximately 2000°C (the temperature of the plasma plume is between 5000—7000°C) [1]. AlterNRG has 5 operating commercial scale facilities of which only 3 process MSW. These plants are located in China and Japan and range in capacity from approximately 2—5 200 tons per day (tpd) [8].

Tees Valley 1 & 2 were facilities that were under construction in the United Kingdom and were to use the AlterNRG technology to convert a targeted 1000 tpd of MSW per facility into electricity. Construction of the first facility was completed but construction of Tees Valley 2 was suspended in November 2015 and the developer, Air Products, publicly announced in April 2016 that it would pull out of the Tees Valley project and the waste-to-energy sector in general. The Tees Valley project experienced technical challenges; it should be noted that it was designed at a larger processing capacity than AlterNRG's currently operating facilities [9].

Table 4.5 lists AlterNRG's operational facilities that process MSW and their annual capacities.

4.2.4.2 InEnTec

InEnTec is a U.S. company that combines plasma gasification and glass melting technology to convert shredded MSW into syngas and slag. The process consists of a downdraft pre-gasifier, a plasma-enhanced melter (PEM), and a thermal residence chamber (TRC). The syngas, which is generated in the PEM via steam gasification, is held in the TRC for approximately 2 seconds at high temperature to allow for full processing of any remaining organic material [10].

Table 4.5 AlterNRG MSW gasification facilities

Plant location	Commissioned in	Capacity (ton/day)
Wuhan,[a] China	2012	150
Utashinai,[b] Japan	2003	220
Mihama-Mikata,[c] Japan	2002	24

[a]Processes biomass.
[b]Processes MSW and autoshredder residue (ASR).
[c]Processes MSW and sludge.

The downdraft gasifier is the heart of the system and does the majority of the conversion. The PEM primarily gasifies residual char in the ash and melts the ash. The TRC is needed to ensure production of a syngas that is free of hydrocarbons. It appears that InEnTec is operating all 3 units in a performance regime that maximizes each unit's efficiency. The InEnTec gasifier technology is a hybrid unit that utilizes each unit operation in a narrow operational regime. This strategy enables each component to operate under optimal conditions where the best performance is obtained for the lowest cost (energy and size). Focusing on the gasifier, it does not attempt to completely convert all the carbon to CO, but instead achieves nearly 85% of the conversion and transfers the remaining 15% to the PEM. Furthermore, in the gasifier, there is no attempt to produce high-quality syngas. The gasifier basically crudely gasifies the fuel to mostly syngas with other gaseous hydrocarbons. This gas mix then enters the TRC where it is used as a polishing step to convert the gaseous hydrocarbons to syngas. This combination seems unique among gasification technology developments. Therefore, no direct comparison can be made to evaluate the InEnTec unit against other commercial units.

InEnTec's system yields a low-to-medium heating value syngas from preshredded feedstock. The produced syngas, which exits the TRC at ambient pressure and a temperature of 1300°C, contains approximately 90% of the energy that is fed to the system in the form of solid waste. Of the energy contained in the syngas, approximately 80% is stored as chemical energy, while 20% is potentially recoverable as thermal energy. Recovery of the thermal energy has not been attempted nor demonstrated. Under the current system design, the facility is capable of processing a variety of feedstocks having heating values similar to most biomass, MSW, and medical waste. Ideally, feedstock with low heating value and high heating value will be blended prior to entering the gasifier.

According to the literature, there are currently 11 InEnTec facilities operating in the U.S. and Asia [3].

4.2.5 Summary table of field scale municipal solid waste gasification facilities worldwide

Waste gasification is a developing industry and there are numerous companies worldwide that operate pilot scale and demonstrational scale facilities. Field scale operations are not abundant due to the economic and technical challenges that are faced in scale up. Table 4.6 provides a summary of field scale waste gasification and pyrolysis facilities worldwide. It

Table 4.6 Summary table of field scale MSW gasification facilities worldwide

Company	Approximate number of field scale operational facilities processing MSW and/or biomass	Approximate capacity range (tons/day)	Locations of facilities
Nippon Steel	33	100–800	Japan, Korea
Thermoselect/ JFE	5	100–600	Japan
Hitachi Zosen	6	30–300	Japan
Mitsui R-21	6	150–400	Japan
Ebara	7	100–400	Japan
Enerkem	1	~300	Canada
Plasco	1	~100	Canada
AlterNRG	2	150–200	China, Japan
Energos	7	100–250	Europe

should be noted that this table does not represent all commercially operating facilities worldwide and only those that are associated with the companies are listed in Table 4.6.

4.2.6 Municipal solid waste gasification for fuel applications

Gasification processes that convert syngas into electricity via gas turbines or to liquid fuels and chemicals via additional downstream processes are not prevalent at field scale operations currently. The main challenge in these gasification applications is that they require syngas clean-up of tars and acid gases which is costly and difficult to manage. Nonetheless, the improved efficiencies of gas turbines and the market demands for liquid fuels and chemicals can make these technologies viable at commercial scale, provided the operations can accommodate fluctuations in the heterogeneous MSW feedstock to consistently yield the designed projected production at high quality.

Enerkem is the current leader and pioneer in field scale operation of waste gasification to generate fuels and chemicals [11]. In 2016, Enerkem began operation of its commercial facility in Edmonton, Canada that gasifies RDF to produce biofuel. Currently, they are producing methanol and in 2017, they plan to also produce ethanol. The Earth Engineering Center at City College of New York (EEC│CCNY) conducted a pilot study at the Enerkem pilot facility in Edmonton to determine the impact that nonrecyclable plastics (NRPs) have on methanol production and process efficiencies and the results are detailed as a case study in the following section.

4.2.7 Case study—Enerkem waste to fuels and chemicals process

Enerkem is a Canadian company that was founded in 2000 and has a Waste to Fuel and Chemicals (WTFC) process that gasifies nonrecyclable, noncompostable MSW to produce biofuels, specifically methanol and ethanol and, optionally, chemicals. Enerkem currently operates a field scale commercial facility on the site of the City of Edmonton's Waste Management Center (EWMC) in Edmonton, Alberta, Canada. The EWMC receives and processes the City of Edmonton's residential and commercial solid waste and sewage biosolids. Enerkem also has a pilot facility located on site of the EWMC in the City of Edmonton's Advanced Energy Research Facility (AERF) and a demonstration facility in Quebec, Canada. The EEC|CCNY conducted a pilot study supported by the American Chemistry Council (ACC) at the Enerkem pilot facility in collaboration with Enerkem and the City of Edmonton. The purpose of the study was to determine the impact of NRP on methanol production and process efficiencies. A description of the Enerkem process and a quantitative summary of the results are provided in this case study.

4.2.7.1 Process description

The Enerkem WTFC process converts MSW into methanol and ethanol via gasification, syngas conditioning, and catalytic syntheses. The feedstock to the Enerkem process is nonrecyclable, noncompostable MSW that is shredded to form RDF fluff. The dimensional size of the fluff ranges from 5 to 9 in. and the typical composition of the RDF that is fed to Enerkem's commercial facility in Edmonton is 60%–70% biogenic material and 30%–40% plastics by weight. This RDF is generated from the material recovery facility (MRF) at the site of the EWMC. The RDF fluff enters the Enerkem gasifier via a lock hopper system that is sequenced to prevent gases from the gasifier to enter the feed hopper or be released into the atmosphere, as well as to minimize the amount of air introduced into the gasifier.

The Enerkem process uses a bubbling fluidized-bed gasifier with a sand-like heat transfer medium. Steam and oxygen (O_2) are injected into the gasifier in a staged fashion to convert the waste to syngas. A unique design feature is the addition of a small amount of carbon dioxide (CO_2) to the gasifier as a purge gas. The purpose of the purge gas is to flow countercurrent to the exiting syngas to remove any entrained particulates, such as sand or unreacted waste. The Enerkem gasifier is typically operated in a temperature range of 700−900°C at a pressure range between 1 and 5 bar. Enerkem has historically targeted a 90%−92% carbon conversion in the gasifier with the remaining 8% of carbon going to char. Fig. 4.1 is a simple process flow diagram of the Enerkem WTFC process. In Fig. 4.1, GSR stands for Gasifier Solid Residues.

The resulting syngas from the gasifier is primarily composed of CO, H_2, CH_4, and CO_2. The syngas exits the top of the gasifier at a pressure of 1−5 bar, subsequently nonactive components are removed, and remaining gas is staged compressed to approximately 1000 psi and reacted with commercial catalyst for methanol production.

The final products of the Enerkem process are methanol or ethanol, depending on the system configuration. Enerkem is also actively involved in the conversion of syngas (via methanol or directly) to yield additional downstream end-products such as dimethyl ether and derived olefins, C2 and C3 acids, and bio-syndiesel.

The Enerkem commercial facility in Edmonton is designed to process approximately 100,000 dry metric tons of nonrecyclable MSW annually and generate ∼ 10 million gallons of biofuel, which equates to approximately 100 gallons of biofuels per dry metric ton of MSW. In a 25-year agreement with the City of Edmonton, Enerkem's affiliate, Enerkem Alberta Biofuels, operates the commercial facility and sells the biofuel product generated from the MSW. The City of Edmonton claims that it will be able to divert up to an additional 30% of its MSW from landfills with the commercial facility thus achieving an overall diversion rate of 90%. The Enerkem commercial facility in Edmonton is shown in Fig. 4.2.

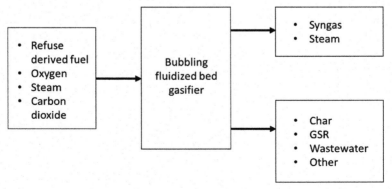

Figure 4.1 Simple process flow diagram of Enerkem process.

Figure 4.2 Enerkem's WTFC commercial facility in Edmonton.

4.2.7.2 Pilot study to determine impact of plastics in the feedstock for gasification

The EEC|CCNY conducted a study at Enerkem's pilot facility to determine the impact that NRP has on methanol production and performance efficiencies of this process. The pilot facility has an operating capacity of approximately 8 tons/day. Feedstock mixtures containing 100% biomass (which in this case was construction and demolition wood chips), 8%, 15%, and 50% NRP by mass were fed into the pilot scale gasifier, and operational performance data were measured and collected, which included syngas composition, gasifier operating conditions, and char generation. The NRP fraction of the feedstock mixtures came from the EWMC's MRF residual and consisted of a 50/50 mixture by weight of rigid and film plastics. Fig. 4.3 shows an example of the NRP feedstock mixture that was tested and Fig. 4.4 shows the Enerkem pilot facility where this study was conducted.

Importantly, the purpose of this study was focused on the impact that plastics have on syngas composition effluent from a gasifier that can be ultimately converted to downstream products such as methanol or ethanol. The Enerkem pilot facility was chosen because of the extensive track record of operation and the insight that it can provide toward a large-scale gasification process. The performance obtained during this specific test

Figure 4.3 Example of NRP feedstock mixture tested in Enerkem pilot study.

Figure 4.4 Enerkem's WTFC pilot facility in Edmonton.

should not be considered representative of the commercial scale operation of the current Enerkem facility. Instead it should be interpreted as a set of comparative tests providing a general correlation to the impact that nonrecyclable plastics would have if they were introduced into a gasifier that was primarily operating on biomass residual waste stream. The reader is referred to U.S. patents #8192647, *"Production of synthesis gas through controlled oxidation of biomass"*; #8137655, *"Production and conditioning of synthesis gas obtained from biomass"* and #8080693, *"Production of ethanol from methanol"* for performance data that are representative of the commercial Enerkem facility.

4.2.7.3 Impact of plastics in the performance from operations data of pilot study
4.2.7.3.1 Syngas composition
Table 4.7 shows the normalized measured syngas compositions and calculated higher heating values of the syngas produced from the various NRP feedstock mixtures that were tested at the Enerkem pilot facility. The chemicals species reported are only the ones that are expected to participate in a syngas to methanol conversion similar to the Enerkem system.

It should be noted that the higher heating values reported in Table 4.7 include small amounts of methane with the BTEX stream being a negligible contribution to the heating value.

4.2.7.3.2 Final product yield and char generation
The methanol production for the trials was calculated based on measured syngas composition, assuming that all H_2 and CO in syngas form

Table 4.7 Normalized syngas composition for plastic feedstock mixtures from Enerkem pilot unit (Normalized gas composition excludes nitrogen, CO_2, and other gas components)

	100% Biomass (wood chips)	92% Biomass/ 8% plastics mixture	85% Biomass/ 15% plastics mixture	50% Biomass/ 50% plastics mixture
H_2	27.76	30.19	26.74	23.85
CO	32.94	34.26	36.86	33.00
H_2O	30.69	22.98	21.81	20.32
CH_4	8.62	10.87	12.31	17.68
C_2H_4	0.00	1.71	2.22	4.31
C_2H_6	0.00	0.00	0.05	0.41
C_3H_8	0.00	0.00	0.02	0.42
Total	100	100	100	100
H/C ratio	0.97	1.13	1.14	1.21
Higher heating value (Btu/cf)	121.2	155.2	231.5	219.1

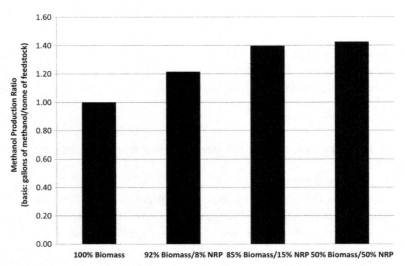

Figure 4.5 Ratio of methanol production for NRP feedstock mixtures compared with 100% biomass feedstock.

methanol at stoichiometric conditions with no additional Water Gas Shift to turn some CO into additional H_2.

Fig. 4.5 shows the ratio of methanol production from the NRP feedstock mixtures to 100% biomass. The results based on the measured syngas composition are shown in the black bars. Several operational factors and the pilot configuration itself contribute to lower yields at the pilot than would be expected in a commercial configuration optimized for methanol production. However, the data does show the tendency toward higher methanol yields as plastics are added. In addition to the plastic C/H addition, actual test operating conditions also affect the final product yields.

Equilibrium modeling was then used to predict syngas composition based on the feedstock mixture composition followed by methanol synthesis. The equilibrium results showed that an increase in NRPs from 15% to 50% would increase the methanol production by 44%, which is expected due to the increased hydrogen to carbon ratio in the feedstock as a result of more plastics.

Measured char production from the pilot study ranged from 18 kg/h for pure biomass feedstock to 4.3 kg/h for feedstock containing 50% biomass/50% plastics. Again, char production was seen to have been affected by a combination of both process conditions and feedstock composition.

Table 4.8 Ratios carbon conversion and energy efficiencies using 100% biomass as a baseline for the pilot unit located at the Enerkem facility

Feedstock	Ratio of carbon conversion of pilot tests (%)	Ratio of energy efficiency of pilot tests (%)
92% Biomass/8% plastics	0.99	1.11
85% Biomass/15% plastics	1.44	2.01
50% Biomass/50% plastics	1.32	1.58

4.2.7.3.3 Carbon conversion and energy efficiency

Table 4.8 shows the calculated ratios of carbon conversion efficiencies and energy efficiencies for different NRP feedstock mixtures that were tested at the Enerkem pilot facility. The ratios are defined taking 100% biomass as a basis. Carbon conversion was defined as the ratio of carbon in the syngas to the total carbon input to the gasifier. The energy efficiency around the gasifier was defined similarly, using 100% biomass as a basis, as the ratio of the energy in the syngas to the total energy input. Again, higher carbon conversion and process efficiencies are seen when using biomass feedstocks that also contain plastics, but the results are also greatly impacted by operating conditions.

With recent technology changes, future Enerkem facilities will target 97% conversion of carbon in the gasifier to syngas with the elimination of char and other organics from the stream. The results of the pilot trials indicate that increased NRP in the feedstock may assist Enerkem in achieving their conversion target as well as improve energy content in the syngas and yield of final product. The results of this study suggest that NRP in the feedstock would enhance overall performance if introduced with the residual MSW stream being fed to the gasifier.

4.3 NONRECYCLED PLASTICS PYROLYSIS FACILITIES

4.3.1 Overview

Nonrecycled plastics (NRP) refer to the plastics in MSW that are not recycled either because of technology and market limitations or because of low recycling rates of waste generators. NRP, which consists of rigid and film plastics, is currently predominantly landfilled despite the high calorific value of plastics that could be recovered through thermal conversion processes. There is a niche market for NRP that is currently developing through the use of pyrolysis technologies. Many pyrolysis technologies are focusing on converting plastics-to-oil that can be sold back into the market for heating or fuel purposes. Similar to gasification of MSW, there are currently very few field scale operations of commercial pyrolysis facilities. One of the major challenges that this technology encounters is related to the economics, specifically, the market price of fossil fuels and heating oils. Pyrolysis requires a more refined feed stream compared to mass burn combustion in order to meet a consistent product quality and it also requires preprocessing of feedstock, which is an added cost. The separation and transport of NRP is an additional cost to municipalities; therefore, an improved waste collection and separation framework is key to improving the economic viability of pyrolysis technologies.

Nonetheless, there is a growing interest in plastics pyrolysis technologies because of the significant inherent chemical energy that can be potentially recovered from NRP. The average heating value of NRP is 35.6 MJ/kg and is higher than that of natural gas and coal [12]. A major benefit of waste pyrolysis compared to other thermal treatments is that it occurs in a reducing inert environment. Therefore, metals that are recovered from the char remain in the valuable metal state because they are not oxidized as they are in combustion and most gasification systems.

There are currently no field scale operations of commercial pyrolysis technologies; however, there is a growing abundance of companies that are developing technologies in this area and have pilot and demonstrational scale facilities. Table 4.9 is a list of companies that are paving the way for plastics-to-oil commercialization and currently have operational plastics pyrolysis facilities at pilot or demonstrational scale.

The following section is a case study of a plastics-to-oil pyrolysis technology developed by Golden Renewable Energy (GRE). This study was conducted by the EEC|CCNY as part of an independent due diligenceand included waste feedstock, syngas, and final product characterization, measured performance data, and calculated performance efficiencies of this technology.

4.3.2 Case study—Golden Renewable Energy

Golden Renewable Energy (GRE) pyrolyzes NRP to produce a diesel and home heating oil that can be sold into the wholesale market. GRE owns and operates a demonstrational scale facility in Yonkers, New York

Table 4.9 Summary table of plastics-to-oil pyrolysis technologies

Company	Locations	Plant capacity	Process description	Feedstock	Final product yield
Agilyx	Oregon, United States	Not specified	Batch	NRP, polystyrene	Claim to yield 1 gallon/ 8.5−10 lb plastic
Vadxx	Ohio	60 tpd[a]	Continuous, rotary kiln	NRP, industrial waste plastic	Claims 0.1 gallons/lb of NRP
Enval	United Kingdom	2000 tons/year (modular)	Microwave pyrolysis	Plastic aluminum laminates	Claims 0.1−0.2 tons aluminum recovered/ year
Golden Renewable Energy	New York	24 tpd	Continuous	NRP, tires	Between 2 and 6 bbl oil/ton of plastic
Pyrocrat	India	3−48 tpd (modular design)		NRP, tires	Claims 0.7−0.9 L/kg plastics
Plastic2Oil	New York		Continuous	NRP	Claims 1 gallon/8.3 lbs of plastic

[a]Commissioned to 16 tpd.

(NY) that has a processing capacity of approximately 8 tpd. The EEC|CCNY conducted a due diligence of the GRE demonstration facility and the results are presented in this case study.

4.3.2.1 Process description

The GRE process accepts all plastic resin types except for polyvinyl chloride (#3-PVC). The NRP feedstock that was processed during the due diligence was material recovery facility (MRF) residual that contained approximately 50% polypropoylene (#5-PP) and 30% high density polyethylene (#2-HDPE). Prior to entering GRE's system, which is referred to as the Renewable Fuel Production (RFP) unit, the NRP feedstock is mechanically shredded on site to 0.75 in.- < 1 in. size, metals are removed with an automated magnet, and paper contaminants are currently removed manually. The expectation during commercial operation would be to automate these operations for cost and safety improvements. The feedstock is transferred by conveyor belt to a surge bin and then transferred to the hopper of the RFP pneumatically.

The NRP feedstock is fed incrementally to the RFP unit during operation controlled by a level sensor in the hopper. The hopper is designed to handle a bulk density of plastic feedstock of approximately 11.3 lb/ft³. The NRP is fed into the extruder of the RFP unit, which operates at approximately 530°C. The purpose of the extruder is to melt the plastics to improve feedstock flow entering the pyrolysis reactors.

The RFP unit has two pyrolysis reactors in series. The first reactor, Reactor 1, volatilizes the plastics into an oil vapor at an approximate operating temperature range of 590–650°C. Plastics that do not volatilize, remain as a solid and fall by gravity into the second reactor, Reactor 2, where they are heated to an operating temperature in the range of approximately 370–650°C. Any solid residual from the RFP unit exiting Reactor 2 is collected in a holding tank. The residual currently is disposed of in an ash monofill but has potential use for material applications.

The gas and vapors generated in the pyrolysis reactors are combined entering a firebox for homogenization. The hot pyrolysis gas (Py-gas) from the firebox is sent through an ethylene glycol cooling jacket to bring down gas temperature to approximately 65°C after which it enters a series of cyclones. The RFP unit has a total of eight cyclones in series that separate out liquid petroleum fractions from the Py-gas. Cyclone 1 separates out the heavies and the final cyclone, Cyclone 8 separates out the most refined fraction, which is similar in composition to diesel. The liquid petroleum fractions are pumped into a single line that goes to an above ground storage tank. The mixed petroleum product from the demonstrational scale RFP unit is currently sold for residential heating purposes. Fig. 4.6 is a simple process flow diagram of the GRE process.

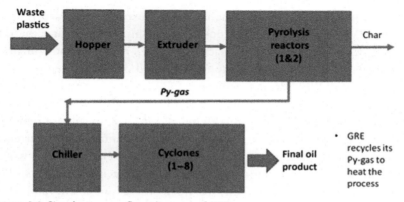

Figure 4.6 Simple process flow diagram of GRE process.

Electricity and natural gas are used to heat the RFP unit during start-up. Approximately 17 kW are required to heat Reactor 2 and all other units in the GRE process are heated via natural gas. During steady-state operations, the Py-gas from the end of the cyclone series is recycled and used to heat the entire RFP unit, with the exception of one burner of the extruder that runs on natural gas throughout the processing period.

The final product of the GRE process is consumer-ready fuels for transport and heating that can be sold into the wholesale market.

4.3.2.2 Due diligence testing

The EEC|CCNY conducted due diligence testing at the GRE demon-strational facility in Yonkers, NY. Independent continuous measurements were taken on-site of Py-gas composition and stack gas emissions, and operating conditions were provided by the GRE engineers. Samples of the condensate fractions including the final product, char, and the waste feedstock were taken and characterized at the EEC|CCNY labs using several analytical techniques including calorimetry and scanning electron microscopy (SEM). Efficiencies of the GRE process were calculated based on the independent collected data and are presented in the following sec-tion. Results of the stack gas emissions testing are presented in Chapter 5, Emissions, of this textbook.

4.3.2.3 Process efficiencies based on operations data from due diligence

4.3.2.3.1 Syngas composition

The Py-gas in the GRE process is cooled and sent through a series of cyclones to condense out oil fractions. The remaining Py-gas after the cyclone stage is recycled back into the auxiliary system of the GRE pro-cess to heat the reactors. Table 4.10 shows the average measured recycled Py-gas composition from the due diligence testing and the calculated Py-gas energy density.

4.3.2.3.2 Product yields and environmental impact

The final product of the GRE process is liquid fuel. The process produces a heavy fraction similar to that of home heating oil but the targeted fuel product is a transportation fuel similar to that of diesel. Based on the due diligence test analysis, the yield of the GRE process was approximately 2 barrels of oil per ton of plastic waste. The final product was characterized

Table 4.10 Composition and heating value of GRE Py-gas

Species	Mol%
H_2	8.12
CO_2	15.50
C_2H_4	17.06
C_2H_6	17.23
C_2H_2	0.02
O_2	0.42
N_2	4.33
CH_4	31.09
CO	6.21
Total	100
Higher heating value (Btu/scf)	927

Table 4.11 Carbon conversion and thermal efficiency of GRE process

Feedstock	Carbon conversion of GRE process (%)	Energy efficiency of GRE process (%)
NRP	83	80

and determined to have qualities similar to that of diesel and have a higher heating value of 15,793 Btu/lb.

Char is a by-product of the GRE process and yields approximately 11 kg of char per ton of plastic waste (based on the due dilgence results). The char was analyzed to have 30% carbon content and contained metals such as titanium (Ti) and iron (Fe). Further information regarding char composition and the air emissions results of the stack testing are discussed in Chapter 5, Emissions, of this textbook.

4.3.2.3.3 Carbon conversion and energy efficiency

Mass and energy balances were performed on the GRE process based on the collected performance data and analysis of the feedstock, product, and residual. Table 4.11 shows the calculated carbon conversion and energy efficiencies. The carbon conversion was defined as the ratio of carbon in the final oil product and the Py-gas to the total carbon input to the pyrolysis unit. The energy efficiency of the pyrolysis process was defined as the ratio of the energy in the Py-gas and the final oil product to the total energy input.

The overall assessment of the GRE process based on the due diligence was that it is able to continuously process NRP to produce a fuel oil comparable in quality to diesel at high performance efficiencies.

4.4 WET SOLID WASTE AND SLUDGE FACILITIES

4.4.1 Overview

High moisture content waste is a difficult waste to manage because it is energy intensive to process. Additional energy must be provided to evaporate the water therefore, most waste combustion and thermal treatment technologies limit moisture content of accepted waste feedstock at a maximum of 25% by weight. Sludge is an organic wet waste that is a byproduct of sewage treatment. Sludge and agricultural wet waste can pose environmental issues because they can contain hazardous contaminants that can leach into the groundwater if left untreated. Thermal treatment systems such as gasification and pyrolysis can process sludge and thermally convert it to syngas. In this context, the technologies are primarily used as a "waste reduction" solutions rather than energy production alternatives because the high moisture content of the waste reduces overall energy efficiencies of the systems.

The following section is a case study of a wet waste gasification technology ,developed by a company called Sustainable Waste Power Systems (SWPS), that was independently analyzed by EEC|CCNY. Operational data and performance efficiencies are provided from extensive testing of both the SWPS prototype and the demonstrational facility.

4.4.2 Case study—Sustainable Waste Power Systems

Sustainable Waste Power Systems (SWPS) is a New York-based company that gasifies high moisture content organic waste to produce heat and electricity. SWPS operates a demonstrational scale facility in Saugerties, NY that has a processing capacity of up to 15 wet tpd. SWPS also has an approximately 1 tpd prototype unit that was donated to EEC|CCNY and testing and analyses are conducted on it for different feedstock mixtures. The results of an independent due diligence conducted by EEC|CCNY at the demonstration facility are presented in this case study.

4.4.2.1 Process description

The patented SWPS process is designed to treat wet organic waste with up to 85% moisture content. Feedstocks that the SWPS process accepts include animal waste, biomass residuals, and byproducts from breweries. Feedstock slurry is introduced into the SWPS system via the main pressure pump (MPP) and enters the devolatilization reactor (DVR). Light

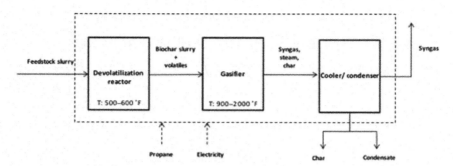

Figure 4.7 Simple process flow diagram of SWPS process.

hydrocarbons in the feedstock slurry are volatilized in the DVR, which is operated between 260–315 °C and at a pressure of approximately 1500 pounds per square inch (psi). The entrained volatile gases and the remaining slurry that does not volatilize in the DVR, referred to by SWPS as the Bio-Char slurry, flow from the DVR and enter the gasifier via the main control valve (MCV). The gasifier operates within a temperature range of 480–1100 °C and is designed to be heated by the SWPS syngas, which SWPS refers to as SynFuel. The last phase of the SWPS system is the cooler/condenser phase which is broken down into three condenser stages. The outputs of the cooler/condenser phase are the syngas, char slurry, and condensate. Fig. 4.7 shows a simple process flow diagram of the SWPS system.

In commercial operation, the SynFuel will be burned for heat and a portion of the SynFuel will be recycled back to the burners of the SWPS system to heat the DVR and the gasifier. The additional products of the SWPS system are the char slurry and the condensate. The condensate is recycled back to the SWPS system to be used in the ancillary heating and cooling system. The char is a byproduct of the process and is currently disposed, but it may have potential for beneficial reuse in terms of metals recovery and possible reuse as a fertilizer.

4.4.2.2 Due diligence testing

The EEC|CCNY conducted due diligence testing at the SWPS demonstrational facility in Saugerties, NY. The feedstock tested during this due diligence was a slurry containing approximately 15%, by weight, cracked corn. Independent continuous measurements of SynFuel composition were taken on-site and the operating conditions were provided by the

Table 4.12 Composition and heating value of SWPS SynFuel

Species	Mol%
H_2	30.37
CO_2	34.77
C_2H_4	5.18
C_2H_6	0.32
C_2H_2	0.45
O_2	0.91
N_2	5.55
CH_4	11.25
CO	11.19
Total	100
Higher heating value (Btu/scf)	345

SWPS engineers. Since the primary purpose of this test was to characterize the SynFuel product, the burner to the gasifier was fueled by propane for the duration of the test instead of recycled SynFuel. Samples of the char slurry and condensate were collected and characterized at the EEC | CCNY labs. Efficiencies of the SWPS process were calculated based on the independent collected data and are presented in the following section.

4.4.2.3 Process efficiencies based on operations data from due diligence
4.4.2.3.1 Syngas composition
The SynFuel in the SWPS process is converted to electricity and a portion of it is recycled to heat the gasifier. Table 4.12 shows the average measured SynFuel composition from the due diligence testing and the calculated SynFuel energy density.

4.4.2.3.2 Product yields and environmental impact
The final product of the SWPS process is SynFuel that is converted to heat and electricity. Based on the results of the due diligence analysis, the SWPS process yields approximately 3200 standard cubic feet (scf) of SynFuel per ton of slurry with an average heating value of approximately 345 Btu/scf. Char is a byproduct of the process and yields approximately 51 lbs per ton of slurry. The char contained approximately 85% carbon by weight. Further information regarding char composition is discussed in Chapter 5, Emissions, of this textbook.

Table 4.13 Carbon conversion and energy efficiency of the SWPS process

Feedstock	Carbon conversion of the SWPS process (%)	Energy efficiency of the SWPS process (%)
Cracked corn slurry (approx. 15% solids)	61.3	67.0

4.4.2.3.3 Carbon conversion and energy efficiency

MASS and energy balances were performed on the SWPS process based on the collected performance data and analysis of the feedstock, product, and residual. Table 4.13 shows the calculated carbon conversion and energy efficiencies. The carbon conversion was defined as the ratio of carbon in the SynFuel to the total carbon input to the gasifier. The energy efficiency of the gasifier was defined as the ratio of the energy in the SynFuel to the total energy input.

Overall, the assessment of the SWPS process was that it can address the wet waste market and reduce the volume of this waste type efficiently.

4.5 CONCLUDING REMARKS ON FUTURE GROWTH OF FIELD SCALE TECHNOLOGY IN WASTE GASIFICATION AND PYROLYSIS

Waste gasification and pyrolysis is still a developing industry when it comes to field scale operations. Most commercial field scale gasifiers are currently in Japan and burn the syngas to produce electricity. Applications of syngas for gas turbines or as an intermediate to produce liquid fuels and chemicals are being explored and Enerkem is the pioneer in this area. Their field scale commercial facility came online in 2016 and is designed to produce biofuel from RDF gasification. Pyrolysis technologies currently operate at the pilot and demonstrational scale and are targeting primarily NRP to produce fuels for transport and heating. Independent analyses of a few waste gasification and pyrolysis technologies indicate that these processes can handle non-recyclable and noncompostable waste, are efficient, yield products similar in quality to virgin fuels and chemicals, and they also meet environmental regulatory standards. The primary challenge that this industry must overcome is making these technologies cost competitive with other waste practices. Operations of waste gasification and pyrolysis can be optimized and costs, consequently, reduced, if the waste feedstock is more homogeneous and if they are used in modular applications. Therefore, the future of commercialization of waste gasification and pyrolysis is largely dependent on the synergy that can be achieved between the front-end collection and the plant processing of waste for these waste conversion technologies.

REFERENCES

[1] WSP, Review of State-of-the-Art Waste-to-Energy Technologies, January 2013.
[2] ISWA, Alternative Waste Conversion Technologies, January 2013.
[3] P. Reddy, Energy Recovery from Municipal Solid Waste by Thermal Conversion Technologies, Talor and Francis Group, Abingdon, UK, 2016.
[4] S. Consonni, F. Vigano, Waste gasification vs. conventional waste-to-energy: a comparative evaluation of two commercial technologies, Waste Manage. 32 (4) (2012) 653–666.
[5] Energos, http://www.energos.com/our-plants/, 2017.
[6] Ebara, http://www.eep.ebara.com/en/products/gas.html, 2017.
[7] Y. Sumio, et al., Thermoselect Waste Gasification and Reforming Process. JFE Technical Report No. 3, 2004.
[8] AltterNRG, http://www.alternrg.com/waste_to_energy/projects/, 2017.
[9] M. Clay, Up in the Air. Recycling & Waste World, 2016, http://www.recyclingwasteworld.co.uk/in-depth-article/up-in-the-air/141467/.
[10] InEnTec, http://www.inentec.com, 2017.
[11] Enerkem, http://enerkem.com/facilities/enerkem-alberta-biofuels/, 2017.
[12] D. Tsiamis, M. Castaldi, Determining an Accurate Heating Value of Non-Recycled Plastics, American Chemistry Council, Washington, DC, March 2016.

Emissions

Contents

5.1 INTRODUCTION

Municipal solid waste (MSW) is a pollutant aggregate stream that all of humanity generates. Therefore, it must be dealt with efficiently and responsibly to avoid adverse environmental and health impacts of significant magnitude. Thermal treatment is advantageous for managing nonrecyclable and noncompostable MSW, because it reduces the volume of this pollutant stream while recovering energy and materials from it in the process. Despite its benefits, thermal treatment of MSW is not as prevalent in the United States (U.S.) as it could be, partially due to public opposition that stems from concerns about the environmental impact of energy recovery facilities. Owing to the combination of the MSW composition and the high-temperature oxidative environments, energy recovery facilities generate air emissions of nitrous oxide (N2O), sulfur oxides (SOx), carbon monoxide (CO), particulate matter (PM), dioxins/furans, and acid

Gasification of Waste Materials.
DOI: http://dx.doi.org/10.1016/B978-0-12-812716-2.00005-4

gases. Additional emissions of these processes are char and bottom ash and wastewater in the form of solid and liquid, respectively. Environmental groups that oppose energy recovery facilities express concern about the production of and release to the atmosphere of hazardous air pollutants (HAPs), such as dioxins/furans and mercury (Hg). However, all operational energy recovery facilities are equipped with air pollution control (APC) equipment to ensure that air emissions meet federal Maximum Allowable Control Technology (MACT) regulatory standards and plant processes are highly controlled to optimize process efficiency which in turn reduces solids and liquids emissions generation.

There is a variation in MSW composition on a daily basis and especially with the seasons. Therefore, there is an expected fluctuation in air emissions of these facilities. The question that is currently being debated by regulatory agencies and the industry is how much variation should be allowed in the emissions and how should the operating data from facilities be used to determine new standards. The Environmental Protection Agency (EPA) together with state and local agencies establishes the limits that apply to waste thermal treatment facilities. There is currently an attempt to update the MACT standards on a pollutant by pollutant basis, however, this procedure will require a long time to finalize and implement. Therefore it is imperative that the relevant agencies are contacted periodically to obtain the most updated information.

The milder oxidizing environment of waste gasification and pyrolysis can potentially facilitate reduced air emissions. Gasification and pyrolysis facilities differ from waste-to-energy (WtE) facilities in that they operate in substoichiometric oxidative and inert environments, respectively. These environments yield a less dilute and lower volume flue gas from which it can be easier to remove pollutants. Furthermore the high temperature and pressure of synthesis gas, also known as syngas, generated from waste gasification and pyrolysis can prevent the formation of pollutants such as NO_x, SO_x, hydrogen sulfide (H_2S), and trace volatiles such as Hg [1]. Syngas applications dictate the primary path of emissions release from waste gasification and pyrolysis. If the syngas is burned in a staged combustion process, the primary emissions release is in the flue gas. The flue gas emissions of staged gasification/combustion may be less than the flue gas of WtE facilities, because the homogeneous gas—gas combustion reaction between syngas and oxidant is more efficient than the heterogeneous solid—gas combustion of WtE [2]. If the syngas is to be used in a gas turbine or converted to liquid fuels or chemicals, it must go through a cleaning step and conditioning to remove tars and acid gases which then end up in the wastewater effluent and must be treated prior to disposal.

This chapter explains the emissions of waste gasification and pyrolysis processes and presents performance data of operating facilities. All data is presented in the context of regulatory standards, and emissions control technologies and techniques are discussed. At the conclusion of this chapter the overall environmental impact of MSW thermal treatment is analyzed within the context of the waste management hierarchy through a life cycle assessment (LCA) comparison of all the waste practices, which include composting, digestion, recycling, and landfill.

5.2 AIR EMISSIONS OF WASTE GASIFICATION AND PYROLYSIS PLANTS

5.2.1 Primary pollutants

5.2.1.1 Nitrogen oxides and sulfur oxides

Nitrogen oxides (NO_x) and SO_x are formed in the combustion of MSW. Nitrogen oxide and dioxide (NO and NO_2, respectively) are the primary components of NO_x [3] and NO_x formation is prevalent for temperatures above 1300°C and when oxygen is not a limiting reagent. Sulfur dioxide (SO_2) is the most predominant form of SO_x and is an acid gas that forms sulfuric acid (H_2SO_4) in the presence of water vapor [4]. NO_x contributes to the formation of ozone (O_3) in the atmosphere and includes N_2O which is a greenhouse gas (GHG) [1]. SO_x can produce sulfuric acid mist that can cause acute organ irritation to humans and animals and also result in acid rain [4].

Gasification and pyrolysis take place in mild oxidizing and inert environments, respectively; therefore less NO_x and SO_x are generated during the conversion of the solid fuel compared to combustion. Most elemental nitrogen (N_2) and sulfur (S) in the MSW ends up as H_2S, carbonyl sulfide (COS), N_2, or ammonia (NH_3) in pyrolysis and gasification [5]. NO_x and SO_x will be generated if the syngas is fired in combustion engines in staged gasification/combustion; however, emissions in the flue gas can be less than WtE facilities because the homogenous gas—gas reactions make it much easier to control the combustion and reduce the amount of excess air needed to ensure complete combustion [2]. S and N_2 removal techniques in syngas cleanup are discussed in Section 5.2.2.

5.2.1.2 Carbon monoxide, carbon dioxide, and volatile organic compounds

One of the primary components of syngas from MSW gasification and pyrolysis is CO. CO can be converted either to hydrogen via water—gas shift reaction or to a final chemical product via reactions such as Fischer Tropsch. Levels of carbon dioxide (CO_2) and volatile organic compounds (VOCs), which are carbon-containing compounds, can be reduced with control of temperature, residence time, and excess air in the combustion stage. CO can remain in the flue gas output if it is not oxidized to CO_2 due to the lack of oxidant or low temperatures. CO is an indicator of combustion efficiency where high CO indicates poor combustion. Semi-suspension-fired refuse-derived fuel (RDF) combustion units have generally higher CO levels than mass burn units [4]. Staged gasification/ combustion can have lower emissions of CO than direct MSW combustion due to improved efficiencies from homogeneous phase combustion [6,2].

5.2.1.3 Dioxins/furans

Dioxins/furans are benzene aromatics containing oxygen and chlorine and are classified by the EPA as known human carcinogens [7,4]. Short-term exposure can cause liver complications and chronic exposure can impair the immune system, developing nervous system, endocrine system, and reproductive functions [4]. Dioxins refer to polychlorinated dibenzo-p-dioxins of which there are 75 congeners and furans refer to polychlorinated dibenzofurans of which there are 135 congeners [7]. Dioxins/furans are formed in a temperature regime between 250 and 700°C with maximum formation at 315°C. Hydrogen chloride (HCl) generated in the gasifier or pyrolysis reactor gets oxidized downstream and the resulting Cl gas reacts with the aromatics in the flue gas to form dioxins/furans. The low levels of or absence of oxygen in the flue gas in gasification and pyrolysis result in reduced formation of dioxins/furans as compared to WtE processes [8].

Formation of dioxins/furans is a result of a complex set of competing chemical reactions and mechanisms. In addition to existing as contaminants in the combusted organic material, another major mechanism of dioxin/furan formation is de novo synthesis in which they are formed in the presence of fly ash containing unburnt aromatics and metals that act as catalysts. De novo synthesis occurs in the presence of oxygen and catalysts at temperatures between 250°C and 450°C; this is typically in waste heat boilers or electrostatic precipitators (ESPs). Lastly, dioxins/furans can form from the thermal breakdown and rearrangement of aromatics that result from complete combustion of organic compounds; this occurs under the same conditions as de novo synthesis [4].

In the case of combustion of MSW, studies have correlated high CO concentrations and other indicators of incomplete combustion with high dioxin/furan formation. Fluctuations in CO (CO peaks), high Cl content, and low temperatures in the secondary combustion zone (660°C) have all been correlated to increased levels of dioxin/furan emissions levels [4]. This can also be the case in gasification and pyrolysis processes if the syngas is directly burned; however, dioxin/furan levels may be reduced to due to more complete combustion of the gas compared to solid MSW and due to reduced levels of oxygen in the syngas and to the amount of oxygen needed for combustion of syngas.

Dioxin/furan formation can be limited to well below regulatory levels through adjustment of process conditions, which include combustion temperature, residence time, post-combustion temperature, and with the use of APC equipment [4].

5.2.1.4 Particulate matter and fly ash

PM consists of solid particles and liquid droplets found in flue gas. They are categorized by size as PM_{10}, which are inhalable particles with diameters of 10 μm or smaller, and $PM_{2.5}$ or "fines," which are inhalable particles with diameters of 2.5 μm or smaller [9]. High temperatures and long residence times allow for more complete conversion of organic matter and reduction in particle size. The source of PM can include inorganic, organo-metallic, and unburned substances entrained in the flue gas. RDF units, which are primarily preferred for gasification and pyrolysis processes due to increased homogeneity of fuel feedstock, tend to have higher PM emissions than mass burn due to suspension feeding of the fuel. PM, especially fine PM, is linked to acute and chronic respiratory and cardiopulmonary effects [4].

Fine PM that remains in the flue gas after heat recovery units is referred to as fly ash. Typically, fly ash accounts for about 1%−3% of the waste input by mass on a wet basis [4]. Typical metals found in fly ash are lead (Pb), cadmium (Cd), and Zn [10]. APC equipment for PM and fly ash is discussed in Section 5.2.2.

5.2.1.5 Acid gases, heavy metals, and tars

Hydrochloric acid (HCl) and SO_2 are the primary acid gases formed from the combustion of MSW. Although gasification and pyrolysis do not directly combust MSW, these species can form from the combustion of syngas. Additional acid gases that can be present in the flue gas are hydrogen fluoride (HF), hydrogen bromide (HBr), and sulfur trioxide (SO_3) but these are in lower concentrations compared with HCl and SO_2. The formation of HCl and SO_2 is directly linked to the Cl and S concentration in the MSW and is not affected by the combustion conditions. However, the combustion environment does dictate which S compounds form. Excess oxygen promotes the formation of SO_2 and SO_3 while oxygen deficient conditions produce H_2S, COS, and elemental S. HCl and SO_2 form acids in the presence of water vapor and can cause corrosion to system equipment. Emissions of acid gases can cause sensitive organ irritation in humans and animals and can cause acid rain [4]. The presence of alkaline constituents such as calcium oxide (CaO) in fly ash can partially neutralize some of the acid gases in the flue gas [5].

Heavy metals are arsenic (As), Cd, chromium (Cr), nickel (Ni), cobalt (Co), copper (Cu), manganese (Mn), Pb, and mercury (Hg). These metals are emitted in PM in the form of oxides and chlorides with the exception of Hg which is emitted as a vapor. Heavy metals are carcinogenic and therefore must be removed from the flue gas prior to release to the atmosphere. Electronic waste (e-waste) is a source of Cd and Hg, and presorting prior to combustion can reduce emissions of these heavy metals. Nonetheless these metals are also present in general MSW and can be found in either the fly ash and flue gas or in the bottom ash depending on the volatility of the metal and the combustion operating conditions. High levels of carbon in the fly ash can improve Hg adsorption in the PM air pollutions control device [4]. APC equipment for flue gas cleaning of heavy metals is further discussed in Section 5.2.2.

Tars are liquid aromatic hydrocarbons mixtures that form during gasification. Tars can damage equipment if they deposit on the walls of the reactor and downstream equipment. Mitigation of tar production includes increased gasification temperatures or residence time, increased concentration of oxidant, and catalytic decomposition [5].

5.2.2 Air pollution control systems

APC equipment is installed in waste thermal treatment systems to meet regulatory air emissions standards. APC equipment employed in the syngas cleaning and conditioning stages and flue gas treatment of waste gasification and pyrolysis plants typically include scrubbers, cyclones, fabric filter bags, and ESPs. Additional removal techniques include catalytic reduction, dry sorbent injection, and flue gas condensation.

Scrubbers are used to remove acid gases, such as HCl and SO_2, and alkaline gases from the flue gas. There are several scrubber types: wet, semi-dry, and dry. The differentiating factor between these scrubbers is the environment in which the flue gas is treated. Wet scrubbers dissolve the flue gas in an aqueous washing solution, such as sodium hydroxide (NaOH), and remove HCl and HF in the first stage and SO_2 in the second stage at a pH close to neutral or alkaline. Semi-dry scrubbers spray the flue gas with an aqueous absorption agent as opposed to saturating it, and dry scrubbers pass the flue gas through an agent that is in the form of a fine, dry powder, such as lime. The acids react with the powder and the solid reaction products are filtered out. The advantage of wet scrubbers is that they are generally less expensive than semi-dry and dry scrubbers; however, the disadvantage is that they produce wastewater that must be treated and discharged. Semi-dry and dry scrubbers are less prone to corrosion and do not generate wastewater; however, they consume more chemicals and yield solid residues [4].

Cyclones, fabric filter bags, and ESP remove PM and fly ash from the flue gas. Cyclones consist of a cylindrical chamber and remove PM from the flue gas via inertia. Owing to their limited removal efficiency, they are employed in facilities as a complement to other flue gas treatment stages. Fabric filter bags are the most commonly employed equipment because they have a high particulate removal efficiency for large and small particles. The particulates are captured in the series of filter bags as the flue gas passes through them and then air is blown in the opposite direction to clean the filters and collect the dust. ESP uses a high electrical voltage to charge particles and deposit them on an electrode for collection. ESP has a limited efficiency in removing small particles and is often installed together with fabric filter bags to meet stringent emissions limits [4].

Catalytic reduction, activated carbon, and dry sorbent injection are employed to reduce NO_x and remove heavy metals, dioxins/furans, and PM, respectively. NO_x can be destroyed via selective noncatalytic reduction (SNCR) or selective catalytic reduction (SCR). SNCR uses dry urea, NH_3, as a reducing agent at temperatures between 900 and 1050°C to form water and N_2 while SCR uses a catalyst, such as a mixture of NH_3 and air, to form oxygen and water. SNCR is cheaper and yields less corrosion problems while SCR is more efficient, achieving reduction rates of up to 85%. Activated carbon and alkali sorbents such as lime and NaOH are injected into flue gas to remove volatile heavy metals and acid gases [4].

Alkali metals are not part of the criteria pollutants but they are closely correlated with PM emissions. Alkalis that contribute to slagging in waste thermal treatment systems are potassium (K), Na, Cl, and silica. The alkali vapors can chemically degrade the efficiency of ceramic barrier filters used to remove PM in hot syngas cleaning systems. Alkali vapors are addressed through cooling of the syngas prior to use in turbines. The energy loss associated with this has led to the investigation of alternatives which include the use of ceramic filters, called "getter beds", that can tolerate high temperatures [5].

In general, gasification and pyrolysis produce a lower volume of syngas than WtE facilities; therefore, smaller APC equipment is required in these facilities. Furthermore, since the syngas is less dilute than flue gas from WtE (due to less N_2 from lack of excess air), it is easier to remove pollutants from the syngas stream during the syngas cleaning stages of gasification and pyrolysis processes. The higher temperature and pressure of the syngas compared to the WtE flue gas also allows for easier removal of pollutants such as heavy metals [5].

5.2.3 Air emissions regulatory standards and performance data

5.2.3.1 Criteria pollutant standards for thermal waste systems

In the U.S., the primary pollutants of concern from waste thermal treatment systems are outlined in the EPA's Clean Air Act. Section 111 of the Clean Air Act regulates emissions of the following nine air pollutants from WtE plants: PM, CO, dioxins/furans, SO_2, NO_x, HCl, Pb, Hg, and Cd. Criteria pollutants of the Clean Air Act are PM, O_3, SO_2, NO_2, CO, and Pb. The National Ambient Air Quality Standards (NAAQS) for the criteria pollutants are provided in Table 5.1.

5.2.3.2 Hazardous air pollutants standards, new source performance standards, and emissions guidelines for thermal waste systems

HAPs are air toxics that are known carcinogens and can cause other serious health impacts. Currently, 187 air pollutants fall under the HAPs standards category and include dioxin/furans, HCl, and H_2S. The Clean Air Act directs EPA to limit HAPs emissions for each industry based on best controlled and lower-emitting sources by the industry. These are known as the

Table 5.1 NAAQS for criteria pollutants

Criteria pollutant	NAAQS
Ozone	70 ppb
Particulate matter	$PM_{2.5}$ (fines) is 12 $\mu g/m^3$—annually
	$PM_{2.5}$ is 35 $\mu g/m^3$—24 h
	PM_{10} (coarse) is 150 $\mu g/m^3$—24 h
Carbon monoxide	9 ppm over 8 h, 35 ppm over 1 h
Lead	0.15 $\mu g/m^3$ in total suspended particles as a 3-month average
Sulfur dioxide	Primary standard: 75 ppb for 1 h
	Secondary standard: 500 ppb average over 3 h, not to be exceeded more than once per year
Nitrogen dioxide	Primary standard: 100 ppb for 1 h
	Secondary standard: annual mean of 53 ppb

MACT standards. Within 8 years of setting these technology standards, EPA has to assess if the HAPs protect public health with an ample margin of safety and protect against adverse environmental effects [11].

The EPA regulates emissions from stationary sources of pollution, which includes MSW combustors, through the National Emissions Standards for Hazardous Air Pollutants (NESHAP), new source performance standards (NSPS), and emissions guidelines (EG) for criteria pollutants and other air pollutants [11].

MSW combustors fall under different categories of regulations of the Clean Air Act based on size of facility and year of construction. The categories are: Large Municipal Waste Combustors (LMWC), Small Municipal Waste Combustors (SMWC), and Other Solid Waste Incineration (OSWI). These all fall under Section 129 of the Clean Air Act. This regulation specifies the limiting emissions of PM, CO, dioxins/furans, SO_2, NO_x, HCl, Pb, Hg, and Cd. Section 129 includes NSPS and EG for MSW combustors and are outlined in the Code of Federal Regulations (CFR) in 40 CFR Part 60.

Table 5.2 defines the MSW combustor categories identified under the Clean Air Act and provides the emissions limits of criteria pollutants and HAPs for each category as provided in 40 CFR Part 60.

5.2.3.3 Application of regulation standards to waste gasifiers and pyrolysis plants

There is no official designation of standards solely for MSW gasification and pyrolysis plants in the Clean Air Act. This is largely due to the fact

Table 5.2 Emissions limits for municipal waste combustors based on 40 CFR Part 60

MSW combustor category[a]	Capacity (tpd, tons per day)	40 CFR Part 60
Large Municipal Waste Combustors (LMWC)	>250	Subpart Cb—emissions limits for CO and NO_x for facilities constructed on or before September 20, 1994 Subpart Ea—emissions limits for CO for facilities for which construction commenced after December 20, 1989 and on or before September 20, 1994 Subpart Eb—emissions limits for CO for facilities for which construction commenced after September 20, 1994 or for which modification or reconstruction is commenced after June 19, 1996 Subpart FFF—federal plan emissions limits for NO_x and CO and compliance schedules and exceptions for facilities constructed on or before September 20, 1994
Small Municipal Waste Combustors (SMWC)	35–250	Subpart AAAA—emissions limits for NO_x, SO_2, CO, PM, dioxins/furans, metals, acid gases, and fugitive ash for facilities for which construction is commenced after August 30, 1999 or for which modification or reconstruction is commenced after June 6, 2001 Subpart BBBB—emissions limits for NO_x, SO_2, CO, PM, dioxins/furans, metals, acid gases, and fugitive ash for facilities constructed on or before August 30, 1999 Subpart JJJ—federal plan emissions limits for NO_x, SO_2, CO, PM, dioxins/furans, metals, acid gases, and fugitive ash and compliance schedules and exceptions for facilities constructed on or before August 30, 1999
Other Solid Waste Incineration (OSWI) 1. Very Small Municipal Waste Combustors (VSMWC) 2. Institutional Waste Incineration (IWI)	<35	Subpart EEEE—emissions limits for NO_x, SO_2, CO, PM, dioxins/furans, metals, and acid gases for facilities for which construction commenced after December 9, 2004 or for which modification or reconstruction is commenced on or after June 16, 2006 Subpart FFFF—emissions limits for NO_x, SO_2, CO, PM, dioxins/furans, metals, and acid gases for facilities that commenced construction on or before December 9, 2004

[a]Although not listed here, it should be noted that the 40 CFR Part 260 also includes emissions limits for hazardous waste combustors, hospital/medical/infectious waste incinerators, and sewage sludge incinerators.

that there is not a significant amount of large scale commercially operating plants of this type currently in the U.S. to establish a typical baseline performance. It should be noted that while the EPA puts forth the federal standards for air emissions, states can provide their own State Implementation Program (SIP) that can establish more restrictive standards in the state than the federal limits. Pilot scale gasification and pyrolysis facilities in the U.S. must go through the appropriate regulatory channels and procedures that are set forth in the SIP [12].

In the European Union (EU), all WtE facilities must comply with the Waste Incineration Directive (WID) 2000 and the Industrial Emissions Directive (IED). The IED sets key requirements for the operation of waste thermal treatment systems, which includes that MSW be combusted at a minimum temperature of 850°C for at least 2 seconds and that the resulting bottom ashes and slag has a total organic carbon content of less than 3%. The IED sets emissions limits for SO_2, NO_x, HCl, HF, total organic carbon, CO, dust, heavy metals, and dioxins/furans [13].

Table 5.3 shows performance data of commercially operating gasification plants in Europe and Japan and compares them to the emissions standards. The performance data is divided into three different data sets based on the literature source and the flue gas basis for the emissions measurements. The regulation standards listed are those that are applicable to the region where the given facility is located. It should be noted that in the last data set, all performance data is compared to the best available technology (BAT) standards listed in the last row of Table 5.3.

5.2.4 Case study: air emissions of a plastics-to-oil pyrolysis facility

The Earth Engineering Center at City College of New York (EEC| CCNY) conducted air emissions testing at a pilot scale plastics pyrolysis facility that converts nonrecycled plastics (NRPs) to diesel and home heating oil. The pilot facility has a capacity of approximately 8 tons per day (tpd) and at the time of the stack testing no APC equipment had been installed. The NRP feedstock that was processed during the air emissions test was residual provided by a local material recovery facility (MRF) and consisted predominantly of shredded polypropylene (#5-PP) and high density polyethylene (#2-HDPE).

Stack gases were measured and analyzed for PM, CO, CO_2, NO_x, SO_2, and (VOCs). PM concentrations were measured with a DUSTRAK DRX meter, NO_x and SO_2 concentrations were measured with an

Table 5.3 Emissions performance data from operating commercial gasifiers

Literature source: UC riverside [6] Emission units: (mg/Nm³) @7% O₂	Description	PM	PM standard	HCl	HCl standard	NOₓ	NOₓ standard	SOₓ	SOₓ standard	Hg	Hg standard	Dioxins/ furans (ng/Nm³)	Dioxins/ furans standard
Ebara	Fluidized bed gasification/ash melting 420 tpd industrial/ MSW 5.5 MW electricity Location: Japan	<1.4	15.4	<2.8	126	41	320	<4	225	<0.007	–	7E−05	0.14
JFE/Thermoselect	Pyrolysis and gasification/ syngas engines and boiler 300 tpd MSW 8 MW electricity Location: Japan	<4.7	15.4	11.6	126	–	320	–	225	–	–	0.025	0.14
Mitsui Recycling 21 (R21)	Pyrolysis and gasification/ steam turbine 400 tpd MSW 8.7 MW electricity Location: Japan	<1.0	15.4	55.8	126	82.8	320	25.9	225	–	–	0.0045	0.14
Nippon Steel DMS	High-temperature gasification 200 tpd MSW 2.3 MW electricity Location: Japan	14.1	15.4	<12.5	126	31.2	320	<21.9	225	–	–	0.045	0.14
Plasco Energy	Plasma arc gasification 110 tpd MSW Location: Canada	12.8	14	3.1	14	150	281	26	70	0.0002	14	0.0092	0.14

Literature source: WSP [33] Emissions units: (mg/Nm³) @12% O₂ dry basis

Description	Dust	Dust standard	HCl	HCl standard	NOₓ	NOₓ standard	SO₂	SO₂ standard	CO	CO standard	Dioxins (ng/Nm³ I-TEQᵃ)	Dioxins standard (ng/Nm³ I-TEQ)
JFEᵇ Slagging gasifier 94,200 tpa RDF Location: Fukuyama, Japan	1.1	<11.1	–	–	84.3	<114.0	3.1	<63.4	4.1	<41.6	6E – 05	<0.06
Mitsui Engineering & Shipbuilding (MES) Slagging gasifier 120,000 tpa MSW Location: Toyohashi, Japan	<1	<20	35.8	<65	45.9	<98.2	56.2	<71.4	–	–	0.0032	<0.01
AterNRG Plasma gasification 180 tpd; 50/50 MSW & ASR Location: Utashinai, Japan	<0.01	9ᶜ	22–50	9ᵈ	24–81	180ᵉ	<5.7	45ᶠ	–	–	0.002–0.0098	0.09ᵍ

Literature source: ISWA [34] Emissions units: (mg/Nm³) @11% O₂, dry basis

	Flue gas (Nm³/t)	Dust	HCl	HF	SO₂	NO₂	CO	Hg	Cd + Ti	Dioxins/furans (µg/N m³)
AlterNRG	1.400–2.400	<3	22–39	–	<1–2	62–82	<29	–	–	0.00059–0.00067
Energos	7.894	0.2	3.6	0.02	19.8	42	2	0.003	0.00002	0.001
Ebara	2.952	<1	2	–	<2.8	29.3	–	<0.005	–	5E – 05
Nippon Steel	5.760	6	3	–	0.5	16	5.2	–	–	0.023
Thermoselect	–	0.2	<5	–	–	14	–	–	–	0.0072
BAT	–	1–5	1–8	<1	1–40	40–100	5–30	<0.05	0.005–0.5	0.01–0.1

ᵃInternational Toxic Equivalent
ᵇBased on 11% O2, dry basis
ᶜ⁻ᵍBased on EU criteria.

Enerac (Model 700), and CO, CO_2, and VOCs were measured with an Inficon Micro GC (Model 3000). Air emissions measurements were taken over a testing period of approximately 3 hours and during that time the process was run at two different conditions. The first condition, referred to as "Pyrolysis gas (Py-gas) to Heat," did not condense out oil from the Py-gas but instead recycled all of the Py-gas to the burners. The second condition, referred to as "Py-gas to Oil," condensed out the oil fraction from the Py-gas and recycled the remaining Py-gas to the burners. One burner runs on natural gas throughout the process and the remaining three burners run off recycled Py-gas after start-up. Table 5.4 shows the average measured emissions from the stacks for the scenarios of Py-gas to heat and Py-gas to oil. All emissions presented are based on the assumptions that none of the Py-gas is flared and that the unit operates at 100% availability (24/7 operation throughout the year).

Table 5.5 compares the measured stack gas emissions of the pyrolysis facility to the emissions limits set forth by the local regulatory agency which in this case was the New York State Department of Environmental Conservation (NYSDEC).

The measured stack gases of the plastics pyrolysis facility were within the regulatory standards for PM and NO_x. The emissions generated at the highest concentration were for VOCs, which in this case was the summation of measured concentrations of acetylene (C_2H_2), ethylene (C_2H_4), and ethane (C_2H_6). All presented measured emissions represent an upper limit of emissions for the plastics pyrolysis facility, because no APC equipment was installed at the time of testing, and it was assumed that the facility operates at 100% availability. The results of this test indicate that pyrolysis of plastics can meet air emissions standards and it is anticipated that emissions will be further reduced after the addition of APC equipment.

Table 5.4 Average measured air emissions of plastics pyrolysis facility

Condition	$PM_{2.5}$ (lb/hr)	PM_{10} (lb/hr)	Total PM (lb/hr)	NO_x (lb/day)	SO_2 (lb/day)	CO (lb/day)	VOCs[a] (lb/day)
Py-gas to heat[b]	3.43E − 04	3.44E − 04	3.44E − 04	0.19	0.01	3.24	4.43
Py-gas to oil[c]	1.65E − 04	1.65E − 04	1.65E − 04	6.96	5.26	12.79	86.76

[a]In this case, VOCs represent the summation of the measured concentrations of C_2H_2, C_2H_4, and C_2H_6.
[b]Measurements were taken from two stacks during this condition. One of the burners ran on natural gas and the other ran on recycled Py-gas. VOC measurements for this condition were based on stack results from three burners since an additional burner was run at this condition during VOC testing.
[c]Measurements were taken from two stacks during this condition; both burners ran on recycled Py-gas. An additional stack's measurements for criteria pollutants (NO_x, SO_2, and CO) were run under the condition Py-gas to heat and are included in this average.

Table 5.5 Measured stack gas emissions of plastic pyrolysis facility compared to regulatory standards

	Measured total stack gas emissions of pilot scale plastics-to-oil pyrolysis facility (capacity: 8 tpd)	Regulatory standards
Total PM	1.11E−03	2.4−6.8[a]
NO_x	14.44	15[b]
SO_2	10.58	N.A.[c]
CO	34.01	N.A.
VOCs	126.62	N.A.

[a]DEC Part 212 Permissible Emissions Rate for new source or modification for facilities with process weight ranging from 1000 to 5000 lb/h.
[b]Emissions standard from DEC Part 212 227-2.4g for other combustion installations.
[c]N.A. indicates "Not Applicable." For a facility of this type and scale, DEC could not provide applicable standards for these emissions. The standards for these criteria pollutants will be specific to the operating parameters of the facility.

5.3 SOLID RESIDUAL AND WASTEWATER

5.3.1 Solid emissions

Solid emissions of waste gasification and pyrolysis facilities are char and ash. Char consists of noncombustibles and unburned organic content. Ash is the fraction of char that consists solely of metals and minerals. Typical metals and minerals found in the ash of MSW gasification and pyrolysis are iron (Fe), calcium (Ca), silicon (Si), Ni, and phosphorus (P). Char can be recycled in these process to burn any remaining carbon in the unburned organic content. The ash, referred to as bottom ash, accounts for approximately 10%−30% of total MSW input and is the solid waste product of gasification and pyrolysis plants that must be either disposed or used in alternative applications [4]. Bottom ash is typically disposed of in ash monofills (which are landfills that contain only ash); however, there are increasing applications of ash reuse: as a material substitute in concrete, as a catalyst for high-temperature processes, or as a substrate for microbial fuel cells (MFCs). These applications are discussed in further detail in Section 5.3.1.2.

5.3.1.1 Toxicity characteristic leaching procedure and leaching environmental assessment framework standards

Ash is a solid emission that needs to be regulated to reduce environmental and health impacts because it contains heavy metals which are known carcinogens. The U.S. combines bottom ash and fly ash for final disposal while all other industrial countries dispose of these streams separately. If not treated and disposed of properly in a monofill or a landfill, the heavy metals in ash can leach into and contaminate the groundwater. To prevent leaching, ash must be treated prior to disposal via techniques such as acid washing or high-temperature vitrification. Acid washing removes the heavy metals from the ash while high-temperature vitrification produces a slag that encapsulates the metals and immobilizes them. Vitrification is more prevalent in Japan where the lack of land space requires that the volume of residual generated by gasification plants is minimized and rendered nonreactive.

In the U.S. the environmental standard to test the leaching of ash is the toxicity characteristic leaching procedure (TCLP). In the TCLP, the pH is adjusted to that of landfill conditions and the leachate concentrate is compared to 31 organic chemicals and 8 inorganic chemicals to determine toxicity [14]. If any of the TCLP standards are exceeded, the ash is considered hazardous and must be dealt with appropriately in treatment and disposal. The Leaching Environmental Assessment Framework (LEAF) is a more robust leaching procedure that the US EPA has recently implemented. LEAF is based heavily on European test methods and differs from TCLP in that it focuses on leaching behaviors over a broad range of environment and test conditions with application of the resulting leaching data to specific disposal or use conditions [15]. The four methods of LEAF fall under SW-846 in 40 CFR Part 261.24 and are: Method 1313-pH Dependence, Method 1314-Percolation Column, Method 1315-Mass Transfer Rates, and Method 1316-Batch L/S [16].

5.3.1.2 Beneficial use of ash

The primary method of ash management in the U.S. is disposal in landfills or ash monofills. Beneficial use of ash refers to alternative applications that utilize the ash as an ingredient of a manufacturing process or as an effective substitute for natural or commercial products, in a manner that does not pose threat to human health or to the environment [17]. There is extensive research by the industry to improve metals recovery from the ash, because it contains Fe and precious metals that could be sold back to the market. Currently, the primary beneficial use application of ash is as a material substitute in concrete production. Additional applications that are being explored are the use of ash as a catalyst for high-temperature processes and as a substrate for MFCs.

Char from biomass gasification has been shown to work as a catalyst for high-temperature processes in research conducted by Klinghoffer and Castaldi of the Earth Engineering Center (EEC). The catalytic performance of char is comparable to commercial catalysts and is attributed to its high surface area and the metals content of the ash fraction. Char of poplar wood gasified in a fluidized bed reactor under steam and CO_2 at temperatures of 550, 750, and 920°C for different periods of time had surface areas ranging from 429 to 687 m^2/g, with higher surface areas achieved at higher temperatures and longer residence times, and micropores were observed. Catalytic activity of the char was demonstrated for the decomposition of methane (CH_4), propane (C_3H_8), and toluene (C_7H_8); the latter being a major component of tar found in syngas. The char yielded a lower light off temperature for methane decomposition compared to the commercial Pt/Al_2O_3 catalyst typically used for this type of reaction [18]. The reader is directed to the works of Dr. Naomi Klinghoffer and Dr. Marion Ducousso for further information regarding the application of char from gasification as a catalyst. The application of char as a catalyst would be a less expensive alternative to commercial precious metal catalysts currently used in industry and would divert the solid residual from gasification and pyrolysis plants from landfills.

Another potential application of ash is as a substrate for MFCs. Wastewater contains exoelectrogens (micorganisms that have the abiity to transfer electrons extracellularly) that can generate electricity if colonized on a conductive substrate. MFCs can be utilized to simultaneously sterilize wastewater and produce electricity for small-scale applications. EEC|CCNY collaborated with Livolt, a start-up company, to test Livolt's MFC design in which WtE ash would serve as the substrate. Preliminary tests indicated that the ash submerged in wastewater yielded several properties that would suggest conductive activity including lowered melting point, hydrophilicity, and color change in the ash which could potentially indicate oxidation. The readers of this text is directed to the works of Webster et al. for further information on the application of WtE ash in MFCs [19].

5.3.1.3 Case study: char composition from biomass gasification and plastics pyrolysis

Chars from wet biomass gasification and plastics pyrolysis processes have been analyzed at the EEC|CCNY labs. Analyses include thermogravimetric analysis (TGA) to determine proximate analysis, scanning electron microscopy (SEM) to determine elemental composition of ash, and X-ray diffraction (XRD) to identify crystalline compounds in the ash. Table 5.6 shows the composition ranges of the chars from the gasification and pyrolysis processes and the average measured energy densities.

The feedstock for the plastic-to-oil pyrolysis facility is NRP consisting of predominantly polypropylene (#5-PP) and high density polyethylene (#2-HDPE) in the form of rigids, flexible packaging, and films. The pyrolysis process operates at approximately 300°C and yields diesel and home heating oil as the final products. The ash content of the char residual from the pyrolysis process consists of Fe (2%−14%), Ca (6%−10%), Ti (6%−8%), and Si (1%−2%). Trace levels ($<$1%) of Na, Mg, Al, and K were also detected. Titanium dioxide (TiO_2) and sNaCl were the predominant crystalline compounds detected in the char via XRD analysis at approximately 30% and 15% by mass, respectively. Volatile content in the char samples ranged from 25% to 50% thus indicating that the char could be further processed in the pyrolysis system to convert more of the carbon into Py-gas. The surface area of the char was measured to be approximately 4.371 m^2/g which is significantly lower than that of conventional catalysts. The composition of the char from this process suggests the potential for recovery of metals, such as Ti, and the high heating value and volatile content indicate that the char can be burned to heat the system or reinjected into the process for further conversion.

The results presented in Table 5.6 for wet biomass gasification are char measurements from separate trial runs of processing food waste and animal waste as feedstock. The process gasifies the feedstock at approximately 600°C. The ash content of the char included Fe (0.5%−3%), K ($\sim$1%), Ni (0.2%−2%), and Cr (0.2%−1%). Trace concentrations ($<$1%) of Na, Mg, Si, S, and P were also detected. The high heating value and the

Table 5.6 Characterization of chars from plastics pyrolysis and wet biomass gasification

Element	Plastics pyrolysis	Wet biomass gasification
C	47−71	85−86
O	8−16	6−12
Ash	16−35	3−9
Cl	2−5	$<$1
Average HHV[a] (MJ/kg)	$\sim$25	$\sim$28

[a]Higher heating value.

measured volatile content of >10% suggest that the char can be utilized as an energy source to heat the system in addition to the recycled syngas that is generated from the process.

5.3.2 Wastewater

Wastewater generated from waste gasification and pyrolysis consists of the tars and acid gases removed during syngas cleanup. General wastewater treatment includes separation of light and heavy tars in a coalescer followed by burning for heat or reinjection into the gasification process and dissolved air floatation to remove suspended solids and oils.

5.4 LIFE CYCLE ASSESSMENT COMPARISON OF WASTE MANAGEMENT METHODS

The most sustainable waste management scenario is one that is integrated across appropriate technologies and utilizes all waste practices within the waste management hierarchy. The extent to which each practice is utilized would be driven by the composition of the MSW. As was mentioned previously, thermal treatment of MSW is not as prevalent as other waste practices in the U.S. partially due to misconception regarding its environmental impact.

LCAs are a modeling tool employed to quantify the overall environmental impact of a practice. LCAs are based on an inventory of data related to environmental burdens of a process which include energy consumption for waste procurement and processing and emissions. There have been numerous studies published in the literature of LCAs of MSW thermal treatment compared to other waste management scenarios. Variability in the waste composition, waste management site location, system boundaries, and program software lead to variability in results of LCA studies. However, a general overview of LCA studies suggests that reduced environmental impact is achieved with incorporation of energy recovery and that lifespan and composition of a material will dictate which practice, recycling or energy recovery, is more environmentally beneficial.

Table 5.7 provides an example of an LCA that compares the overall environmental impact of energy recovery to other waste practices. Specifically the study assesses the impact categories for WtE (Scenario 3)

Table 5.7 Comparison of LCA impact categories for waste management scenarios

Scenario	GWP (kt CO_2)	AP (t SO_2)	EP (t NO_3)	Dioxins (g TCDD[a])
Scenario 0—landfill				
Gross	*1914*	546	126	0.24
Net	–	–	–	–
Scenario 1—landfill with biogas recovery				
Gross	966	338	126	0.35
Net	868	186	126	0.29
Scenario 2—MSW sorting with digestion, composting, and RDF combustion				
Gross	704	852	n.a.[b]	0.25
Net	**−340**	**−441**	n.a.[b]	**−0.28**
Scenario 3—WtE				
Gross	948	*1902*	n.a.[b]	*1.38*
Net	224	780	n.a.[b]	0.92

[a]TCDD is tetrachlorodibenzo[p]dioxin.
[b]For these scenarios, landfilled wastes are without significant organic content.
Source: Data from Cherubini et al., Life cycle assessment (LCA) of waste management strategies: landfilling, sorting plant, and incineration, Energy 34(12) (2008) 2116–2123.

compared to scenarios of only landfilling (Scenario 0), landfilling with biogas recovery (Scenario 1), and an MSW sorting plant that sends the organic fraction to anaerobic digestion and composting and converts the inorganic fraction into RDF for combustion (Scenario 2). The LCA is based on the city of Rome in Italy and the waste composition generated there contains approximately 50% organic kitchen waste and 23% plastics. The impact categories assessed in this study are global warming potential (GWP), acidification potential (AP), eutrophication potential (EP), and dioxins. The impacts are global impacts and "gross" refers to the effective emissions released during the life cycle of the practice and does not take into account environmental benefits from energy outputs while "net" does include them. Italicized entries indicate the worst ranking in each impact category and bold entries indicate the best. Based on the results presented in Table 5.7, the best waste management practice for the waste composition presented in this LCA study was the one that employed digestion and composting of the organic fraction and RDF combustion of the inorganic fraction [20].

As was mentioned previously, energy consumption is a factor that is assessed and quantified on a functional unit basis in LCA studies.

Table 5.8 Energy and resource consumption comparison of landfilling and energy recovery of MSW

Impact category	Landfilling	RDF production and combustion	Mass burn combustion
Energy consumption (MJ/kg restwaste[a])	−0.67	−4.95	**−6.35**
Crude oil consumption (g/kg restwaste)	−6.32	−51.9	**−68.4**
Water consumption (g/kg restwaste)	−16.2	**−69.1**	124.7
Occupied landfill volume (m³/t)	1.43	0.49	**0.27**

[a]Restwaste refers to MSW from household collection.
Source: Data from U. Arena, et al., The environmental performance of alternative solid waste managementoption: a life cycle assessment study, Chem. Eng. J. 96 (1−3) (2003) 207−222.

Table 5.8 compares the energy and resource consumption of landfilling, RDF production and combustion, and mass burn combustion. The results presented in Table 5.8 are from an LCA study for the city of Campania in southern Italy and is based on a waste composition of approximately 30% food waste, 23% paper and paperboard, and 10% plastics. Italicized entries indicate the worst ranking in each impact category and bold entries indicate the best. Based on the results in Table 5.8, it is suggested that energy recovery processes generally consume less energy and resources overall compared to landfilling [21].

It is well known that recycling is a key component of any sustainable waste management infrastructure. However, there is often misconception that recycling and energy recovery cannot coexist, because it is believed that WtE would discourage waste generators from recycling. Actually, both waste practices must be employed because not all of the MSW that is generated can be mechanically recycled due to technology, material, and market limitations. A survey of LCA studies on recycling versus other waste management practices for different waste materials was conducted by Bjorklund et al. [22] and an excerpt of the results is presented in Table 5.9. One of the findings from this study was that, in certain cases, the energy savings from recycling of paper and paperboard was not significantly more than from WtE and was dependent on factors such as paper quality, energy source avoided by WtE, and energy source at the paper mill [23].

LCAs are a useful tool to provide preliminary insight into the overall environmental impact of waste practices. However, the variability of waste

Table 5.9 Overview of recycling LCA studies

Reference	Recycled material	Avoided electricity	Avoided heat	Landfill time frame	Total energy	GWP
Arena et al. [21]	PE & PET	Italian mix[a]	n.a.	Short	R < I < L	R < I
	Paper	Coal	n.a.	Not clear	—	R < L
	Glass	n.a.	n.a.	Not clear	—	R < L
	Steel	n.a.	n.a.	Not clear	—	R < L
Craighill and Powell (1996) [24]	Al	n.a.	n.a.	Not clear	—	R < L
	PET	Coal	n.a.	Not clear	—	R < L
	HDPE	Coal	n.a.	Not clear	—	R < L
	PVC	Coal	n.a.	Not clear	—	R < L
Edwards and Schelling (1996) [25]	Al		n.a.	Short	R < L < I	R < L < I
Edwards and Schelling (1999) [26]	Glass	n.a.	n.a.	Short	R < L < I	R < L < I
Eriksson et al. (2005) [27]	Cardboard	Hard coal	Biofuel	Short (100 years)	R < I < L	R = I < L
	Cardboard	Swedish mix[b]	Biofuel	Short (100 years)	R < I < L	R < I < L
	PE	Hard coal	Biofuel	Short (100 years)	R < I < L	R < I < L
	PE	Hard coal	Oil	Short (100 years)	R < I < L	R < I < L
	PE	Swedish mix	Biofuel	Short (100 years)	R < I < L	R < I < L
Finnveden and Ekvall (1998) [28]	Paper packaging	n.a.	Fossil fuel	n.a.	R < I	I < R
	Paper packaging	n.a.	Biofuel	n.a.	R < I	Mixed
	Paper packaging	n.a.	Solid waste	n.a.	R < I	Mixed
	Newspaper	n.a.	Forest residue	Long	R < I < L	R < I < L
	Newspaper	n.a.	Natural gas	Long	R < I < L	R < I < L
	Corrugated cardboard	n.a.	Forest residue	Long	R < I < L	R < I < L
	Mixed cardboard	n.a.	Forest residue	Long	R < I < L	R < I < L
	PE	n.a.	Forest residue	Long	R < I < L	R < I < L

Finnveden et al. (2005) [29]	PP	n.a.	Forest residue	Long	R < I < L	R < I < L
	PS	n.a.	Forest residue	Long	R < I < L	R < I < L
	PVC	n.a.	Forest residue	Long	R < I < L	R = I < L
	PET	n.a.	Forest residue	Long	R < I < L	R < I < L
	PET	n.a.	Natural gas	Long	R < I < L	R < I < L
	PE	n.a.	Forest residue	Long	I < R < L	I < L < R
	PP	n.a.	Forest residue	Long	I < R < L	I < L < R
	PS	n.a.	Forest residue	Long	I < R < L	I < L < R
	PVC	n.a.	Forest residue	Long	I < R < L	I < L < R
	PET	n.a.	Forest residue	Long	I < R < L	I < L < R
	Plastic	n.a.	Oil, natural, gas, coal	Short	—	R < I < L
Mølgaard (1995) [30]	Plastic	n.a.	Oil, natural, gas, coal	Short	—	R < I < L
	Plastic	n.a.	Oil, natural, gas, coal	Short	—	I < L < R
	Plastic	n.a.	Oil, natural, gas, coal	Short	—	I < L < R
Pickin et al. (2002) [31]	Paper	Australian mix[c]	n.a.	Short	—	I < L < R
Wollny et al. (2002) [32]	Plastic	n.a.	n.a.	Short	R < I < L	R < L < I

[a]Italian mix: oil 47%, gas 22%, coal 11%, nuclear 11%, hydro 9%.

[b]Swedish mix: 45.6% hydro, 41.9% nuclear, 6.2% biofuels, 3.3% coal, 1.2% oil, 1.1% natural gas.

[c]Australian mix: not specified, mainly fossil fuels.

Source: Data from Bjorklund et al., Recycling revisited—the life cycle comparison of global warming impact and total energy use of waste management strategies. Resour. Conserv. Recycl. 44(4) (2005) 309–317.

composition, site location, system boundaries and assumptions, and software must be taken into account when interpreting and comparing data from various LCA studies. Along with LCAs, actual measured performance data from the different waste practices should be incorporated in any environmental review by decision-makers in order to provide the most accurate and informed environmental assessment for optimal integrated waste sustainability.

5.4.1 Case study—emissions of cement kiln using refuse-derived fuel compared to petcoke

The previous section compared the environmental impact of thermal conversion to the other waste management practices. This case study compares the environmental impact of waste thermal conversion to conventional energy practices for manufacturing. Thermal conversion of waste can be applied in the cement kiln industry by substituting waste for the traditional fossil fuels that are used to power the kilns. EEC|CCNY conducted a pilot study at the commercial facilities of Cemex, a North American cement manufacturer, to determine the impact that RDF would have on cement kiln emissions if it substituted a portion of the petcoke that is typically used to power the kilns. Pilot testing took place at the Cemex facilities in San Antonio, Texas and Tepeaca in Mexico and the fuel feedstock test mixture contained 16% RDF by weight. The measured emissions were compared to performance data from operation with 100% petcoke, and kinetic modeling was employed to determine evolution of gas species along the cement kiln reactor. Table 5.10 presents the measured emissions of NO_x, SO_2, and CO for the 16% RDF test feedstock compared to the 100% petcoke.

As shown in Table 5.10, RDF substitution reduced cement kiln emissions for NO_x and SO_2. The measured data indicated a 20% increase in

Table 5.10 Measured emissions of RDF substitution in cement kilns compared to petcoke

Fuel feedstock	Criteria pollutant (mg/Nm³)		
	NO_x	SO_2	CO
100% petcoke	766	12.9	295
84% petcoke, 16% RDF	265	8.5	355
Reduction in emissions	65%	34%	−20%
EPA standard, existing kilns	630	830	410
EPA standard, new kilns	200	130	320

CO although the kinetic modeling calculated a 35% reduction in CO. This discrepancy is attributed to the presence of oxygen in the biogenic fraction of RDF which may not get oxidized in the actual process and remain as CO. This is similar to the phenomenon of native NO_x in the MSW feedstock to WTE plants. The reader is directed to the report of Jiao Zhang of EEC at Columbia University for further detail on the emissions comparisons of dioxins/furans, heavy metals, acid gases, and the kinetic modeling results for this study.

5.5 CONCLUDING REMARKS

Emissions of waste gasification and pyrolysis facilities are primarily air emissions of NO_x, SO_2, CO, and PM, and solid and liquid emissions of char and wastewater, respectively. Air emissions are controlled with both process control and APC equipment and performance data of operational facilities show that emissions are well below regulatory standards for both criteria and hazardous pollutants. Wastewater is treated prior to disposal and numerous applications are currently being explored for beneficial reuse of ash to minimize solid residual disposal. Case studies conducted by EEC | CCNY included air emissions testing of a plastics-to-oil pilot facility, char analysis from wet biomass waste gasification and plastics pyrolysis processes, and emissions analysis for RDF substitution in cement kilns. A comparison of LCAs for different waste management scenarios from the literature suggests that energy recovery has less of an overall negative environmental impact than landfilling and in certain areas, it was better than anaerobic digestion. However, it must be noted that there are limitations to LCAs due to variability in assumptions, systems boundaries, and software; therefore, it is recommended that environmental evaluations include measured data as much as possible.

Waste gasification and pyrolysis, in addition to WtE, are primarily waste management methods that are meant to reduce the volume of the pollutant that is MSW. The environmental advantages of these methods are that they save greenfield space and, as an added benefit, they can recover energy and materials in the process. The high-temperature processing of a variable material such as MSW requires the continuous monitoring and assessment of emissions to ensure that regulatory standards for all pollutants of concern are met. The current performance data that has been shared by the industry and has also been collected by independent third parties from waste gasification, pyrolysis, and WtE technologies are below all regulatory standards, therefore suggesting that these practices can be incorporated readily into the waste management infrastructure and have a positive environmental impact.

REFERENCES

[1] National Energy Technology Labs, https://www.netl.doe.gov, 2017.
[2] S. Consonni, F. Vigano, Waste gasification vs. conventional waste-to-energy: a compara-tive evaluation of two commercial technologies, Waste Manag. 32 (4) (2012) 653—666.
[3] EPA, Overview of Greenhouse Gases, https://www.epa.gov/ghgemissions/overview-greenhouse-gases#nitrous-oxide, 2017.
[4] P. Reddy, Energy Recovery from Municipal Solid Waste by Thermal Conversion Technologies, Talor and Francis Group, Park Drive, UK, 2016.
[5] A. Klein, Gasification: An Alternative Process for Energy Recovery and Disposal of Municipal Solid Wastes, Columbia University, New York, NY, 2002.
[6] University of California, Riverside, Evaluation of Emissions from Thermal Conversion Technologies Processing Municipal Solid Wastes and Biomass, 2009.
[7] P. Deriziotis, Substance and Perceptions of Environmental Impacts of Dioxin Emissions, Columbia University, New York, NY, 2004.
[8] N. Klinghoffer, M. Castaldi, Waste to Energy Conversion Technologies, Woodhead Publishing, Sawston, Cambridge, 2013.
[9] EPA, Particulate Matter (PM) Basics, https://www.epa.gov/pm-pollution/particulate-matter-pm-basics, 2017.
[10] G. Tchobanoglous, F. Kreith, Handbook of Solid Waste Management, second ed., McGraw-Hill, New York, NY, 2002.
[11] EPA, Hazardous Air Pollutants, https://www.epa.gov/haps, 2017.
[12] EPA, Clean Air Act Guidelines and Standards for Waste Management, https://www.epa.gov/stationary-sources-air-pollution/clean-air-act-guidelines-and-standards-waste-management, 2017.
[13] Department for Environment, Food, and Rural Affairs, Advanced Thermal Treatment of Municipal Solid Waste, www.defra.gov.uk, February 2012.
[14] R.J. Watts, Hazardous Wastes: Sources, Pathways, Receptors, John Wiley & Sons Inc, Hoboken, NJ, 1998.
[15] D. Kosson, et al., Leaching Test Relationships, Laboratory-to-Field Comparisons and Recommendations for Leaching Evaluation using the Leaching Environmental Assessment Framework (LEAF), USEPA, September 2014.
[16] TestAmerica, An Introduction to U.S. EPA's Next Generation of Leaching Testing. n.d. http://www.testamericainc.com/media/1606/testamerica-leaf.pdf.
[17] USEPA, Definitions: Utilized in the Re-TRAC Connect State Measurement Template. n.d. https://www.epa.gov/sites/production/files/2015-09/documents/smp_definitions.pdf.
[18] N. Klinghoffer, Utilization of Char from Biomass Gasification in Catalytic Applications, Columbia University, New York, NY, 2013.
[19] Webster, et al., Electrical surface modification of waste to energy ash for sustainable energy applications.
[20] F. Cherubini, et al., Life cycle assessment (LCA) of waste management strategies: landfilling, sorting plant, and incineration, Energy 34 (12) (2008) 2116—2123.
[21] U. Arena, et al., The environmental performance of alternative solid waste manage-ment option: a life cycle assessment study, Chem. Eng. J. 96 (1—3) (2003) 207—222.

[22] Bjorklund, et al., Recycling revisited—the life cycle comparison of global warming impact and total energy use of waste management strategies, Resour. Conserv. Recycl. 44 (4) (2005) 309–317.

[23] A. Bjorklund, G. Finnveden, Recycling revisited—the life cycle comparison of global warming impact and total energy use of waste management strategies, Resour. Conserv. Recycl. 44 (4) (2005) 309–317.

[24] A.L. Craighill, J.C. Powell, Lifecyle assessment and economic evaluation of recycling: a case study, Resour. Conserv. Recycl. 17 (1996) 75–96.

[25] D.W. Edwards, J. Schelling, Municipal waste life cycle assessment part 1, and aluminium case study, Trans. IChemE B 74 (1996) 205–222.

[26] D.W. Edwards, J. Schelling, Municipal waste life cycle assessment. Part 2. Transport analysis and glass case study, Trans IChemE B 77 (1999) 259–274.

[27] O. Eriksson, M. Carlsson Reich, B. Frostell, A. Björklund, G. Assefa, J.-O. Sundqvist, et al., Municipal solid waste management from a systems perspective, J. Cleaner Prod. 13 (2005) 241–252.

[28] G. Finnveden, T. Ekvall, Life-cycle assessment as a decision-support tool—the case of recycling versus incineration of paper, Resour. Conserv. Recycl. 24 (1998) 235–256.

[29] G. Finnveden, J. Johansson, P. Lind, A. Moberg, Life cycle assessment of energy from solid waste. Part 1. General methodology and results, J. Cleaner Prod 13 (2005) 213–229.

[30] C. Mølgaard, Environmental impacts from disposal of plastic from municipal solid waste, Resour. Conserv. Recycl. 15 (1995) 51–63.

[31] J.G. Pickin, S.T.S. Yuen, H. Jennings, Waste management options to reduce greenhouse gas emissions from paper in Australia, Atmos. Environ. 36 (2002) 741–752.

[32] W. Wollny, G. Dehoust, U.R. Fritsche, P. Weinem, Comparison of plastic packaging waste options. Feedstock recycling versus energy recovery in Germany, J. Ind. Ecol. 5 (3) (2002) 49–63.

[33] WSP, Review of State-of-the-art Waste-to-Energy Technologies, 2013. http://www.wtert.com.br/home2010/arquivo/noticias_eventos/WSP%20Waste%20to%20Energy%20Technical%20Report%20Stage%20Two.pdf.

[34] ISWA, Alternative Waste Conversion Technologies, January 2013.

Critical Development Needs

Contents

To review a technology and assess its possible success on a commercial scale, the main driving forces of the markets have to be considered. The need for new or alternative fuels is usually regulatory and economically driven, while the development of new sustainable technologies at large scale is also motivated by public interest and government incentives.

Gasification has a series of impediments, and although research in this field has made significant progress, many of the facilities operators still struggle with a few, yet to be solved, challenges. Problems such as gas cleaning, corrosion and fouling in heat exchangers, ash sintering, and tar removal are being discussed in this chapter.

Gasification of municipal solid waste (MSW) is a growing industry due to certain factors. Increased waste generation coupled with decreasing local landfill space and environmental concerns will force municipalities in the United States to consider alternative methods of waste management such as thermal treatment. Currently, approximately 12.9% of MSW (32.8 million tons) in the United States is sent to combustion waste-to-energy (WTE) facilities per year for energy recovery. While WTE is effective, municipalities are showing an interest in gasification and

Gasification of Waste Materials.
DOI: http://dx.doi.org/10.1016/B978-0-12-812716-2.00006-6

pyrolysis technologies, because they occur in reduced oxygen environments, which suggest a higher efficiency compared to traditional incineration, they have the ability to produce a syngas and do not have the public perception issues of incineration. Additionally, there is a speculation that the syngas can be converted to liquid fuels (and chemicals) via a Fischer—Tropsch process. One of the drawbacks of waste gasification is that syngas heating values vary based on the feedstock and efficiency of the thermal process ranging from 300 Btu ft^{-3} for biomass waste to 900 Btu ft^{-3} for nonrecyclable plastic waste with myriad contaminants.

The potential for combining prime mover installations (i.e., gas turbine (GT) and reciprocating engines) with gasification systems is likely the best path forward for the foreseeable future compared to syngas conversion to liquid fuel. The interest in gasification is growing because there is a potential to convert the produced gas (syngas) into a higher value material such as a liquid fuel or commodity chemical. The potential to use waste (both MSW and biomass) as feedstock is immense. There is nearly 10^9 tons of waste landfilled globally on an annual basis. This translates into approximately 10^{13} MJ of thermal energy that could be produced if this waste was diverted to a thermal conversion system.

However, it is unlikely that all waste landfilled today will result in diversion to thermal conversion facilities. The large energy content available in the waste and current public acceptance encourages the recent progress of gasification technology development. However, the two major factors that will limit the use of gasification of wastes to produce chemicals and fuels are the handling/preprocessing of wastes and the syngas cleanup prior to downstream conversion systems.

Any system that requires sorting, shredding, and other manipulations is the first limitation. This will result in size restrictions and an economic disadvantage to systems that can gasify "as-received" wastes. Currently, nearly all MSWs are converted into a refuse-derived fuel (RDF) that is fed into the gasifier. The notable exceptions that process "as-received MSW" are Energos, Thermoselect, and Covanta's Cleergas. RDF has significant benefits that enable gasification technologies such as a higher heating value, better homogeneity in physical and chemical compositions and easier storage, handling, and transportation. On average, 75%—85% of the weight of as-received MSW is converted into RDF, and approximately 80%—90% of the Btu value is retained. It is possible to recover up to a net electrical efficiency of 18%—22% theoretically achievable with a boiler-steam turbine system in an industrial-scale, conventional, incineration plant. Furthermore, a plasma gasifier coupled to a GT-combined cycle power plant can possibly achieve up to 46.2% efficiency.

Prior to the utilization of the produced syngas by a combustion system, it must be cleaned to the engine manufacturer's specifications. Currently, all attempts must cool the gas, reducing the efficiency of the system. Tars, heavy metals, halogens, and alkaline compounds are contaminants in the syngas that can cause operational problems. Until the issue of cooling and cleaning the syngas for catalytic reactor or GT specifications is solved, i.e., how to clean hot syngas to catalytic reactor or combustion engine specifications, all gasification processes will send their product gas to a direct-connect boiler combustion system.

Although gasification of homogeneous fuels has been successful (i.e., coal, biomass, plastic, etc.), the only operating gasification systems for MSW have been done by Thermoselect, Energos, and demonstrated by Covanta's Cleergas. Currently, a large number of uncertainties (regarding performance, reliability, and economics) exist associated with coupling gasification facilities to systems other than direct boiler combustion. Since none of the mentioned waste gasification plants are currently sending the produced syngas to a conversion system, the actual emissions that would be achieved using combustion are unknown. Typically, each developer or technology supplier would prefer to have a reference facility when bidding for projects so it is unknown how many projects will provide confidence to municipalities going forward.

The experience of gasification in Europe is not good. Gasifiers were built in Europe in the 1970s and 1980s, notably Thermoselect; however, due to economic and operational difficulties, they were decommissioned. Since then, conventional combustion systems have become so dominant that there is no real opportunity for other thermal conversion technologies to develop. In the United States, waste gasification technology was introduced in the 1970s and 1980s as well. At that time the MSW quality had a low heating value and contained a significant amount of recyclables such as paper and plastic. However, the experience of operating on such a heterogeneous, low heating value feedstock was not fully developed and gasification systems ultimately failed. Recently, there has been a resurgence in gasification processes that are in development or pilot stages. This resurgence is likely in response to environmentalist and public resistance of incineration. One notable example is the Covanta Cleergas system that combines the expertise and reliability of a conventional WTE operator with an as-received gasification technology, yet there is no interest in developing a site in the United States. Another hurdle for the growth of waste gasification in the United States is the lack of national framework for MSW disposal and the extensive land area in unpopulated regions, which favors landfilling. Finally the eternal hope that recycling and composting will greatly impact the landfill diversion rates, demonstrates the lack of public understanding, and perhaps, interest in the thermal conversion of waste.

Worldwide Japan has the most reliable experience in waste gasification, yet the overall market is small. There are about 30 companies engaged in the development of gasification categorized into (1) fluidized bed, (2) kiln, and (3) direct melt technologies. China is at the stage of vast expansion of all energy conversion systems and has been one of the few to develop circulating fluidized bed gasifiers and combustors for MSW. South America, particularly Brazil and Chile, is exploring the implications of thermal conversion of MSW to energy and chemicals. However with a major focus on biomass, sugarcane bagasse and other perceived environmentally friendly solid fuels the interest in MSW conversion in South America is limited.

Several plants over the course of years have failed due to technical and economic reasons. Stoller and Niessen [1] remind of few such pyrolysis technologies developed in the 70s in the United States, which were not successful or have failed shortly after being put into operation. These technologies were pyrolyzing directly/indirectly heated solid waste in rotary inclined chambers or vertical chambers at low/high temperatures and heated by air (using natural gas) or pure oxygen torch.

The reasons behind failing are many as pyrolysis and gasification technologies encounter many challenges.

6.1 REGULATIONS AND MARKETS

What makes a technology viable are financial incentives, but technologies to produce energy from biomass have received little fiscal and financial incentive compared to other renewable projects such as solar photovoltaic and wind. The cost of the technology is directly related to the cost of feedstock, conversion efficiency, and scale and market value of the products from the process. All these are interconnected and as a technology advances, the cost will change and will become more profitable.

A major role in this is influenced by legislation, which influences the market interest and development capacity.

The motivation behind regulations has mostly been global warming, e.g., decrease CO_2 emissions, but also price and the need to find alternative fuels for transportation. Currently in the United States, there are very few and poorly developed subsides for renewable energy or carbon credits. Examples of successful European practices such as CO_2 tax could lead to a higher use of biomass and development of alternative technologies. Owing to cost effectiveness and availability, biomass-based technologies are not yet competitive to fossil fuels, which contribute to almost 80% of the world primary energy consumption, while only 14% is represented by renewables and 6% by nuclear sources [2].

To evaluate the economic viability of a technology such as biomass into biofuels for example, a series of considerations must be taken into account starting with the cultivation, harvesting, and distribution of the biomass. These include the economic and energy costs for fertilizers use, fuels for machines, and water for crops to the site, preprocessing such as drying and shredding, storage, and the expense of the technology to convert the feedstock into biofuels. To fully assess the success of a potential technology and to determine the impact on the market upon integrating a liquid fuel from biomass technology, a complete study incorporating all aspects must be done. When the proper conditions are created by the policy makers, the question of funds and capital investment comes into attention.

When discussing the MSW management issues in the United States, the municipalities strive to avoid landfills due to liabilities and costs associated with short- and long-term maintenance and the associated regulations. Yet, currently the initial cost of a landfill is lower than installation of a thermal conversion system. Technologies such as waste gasification would bring a few sources of revenue to municipalities ranging from production of fuels and energy to complete recovery of metals and non-ferrous material.

Discussed in the fuel pretreatment section of this chapter the gasification process requires the removal of several types of material from the incoming waste stream for high-quality product gas. Some of these materials are commodity recyclables and are more valuable when sold compared to converted for fuel value. Therefore a private technology developer or a municipality would benefit from recyclables revenue, payments from tipping fees for taking waste, which range from $30 to 100 per ton [3], and sale of valuable products such as liquid fuels and electricity. Prices for commodities have been dropping for more than 5 years, generating low returns for the recycling industry, and causing a number of facilities to close in California, Colorado, Florida, and Missouri. Some of the US largest recyclers, such as Waste Management, sold or closed over 30 recycling facilities over the past few years, while rePlanet announced that is closing 191 centers throughout California [4]. The main advantage of these facilities remains the cost avoided by diverting waste from landfill and transportation associated, but also meeting recycling targets imposed now in many of the states.

Unfortunately the gasification technologies are complex and involve many parts and for this reason many challenges emerge. Based on the Combustion and Catalysis Laboratory (CCL) experience and collaboration with different research groups and industrial sponsors, these challenges would be addressed in detail in this chapter.

6.2 GAS CLEANING

Syngas is the main product of gasification process, and the subsequent uses of this product are very much dependent on its quality and need of additional treatment such as cleaning or upgrading. Different methods for syngas cleaning are commercially available, many of which are used for flue gas cleaning in combustion systems. Many of these gas cleaning methods, such as fabric filters, rotating particle separators, or water scrubbers [5], require a low gas temperature. For an energy efficient system and lower operational costs, hot gas cleaning is done using ceramic filters, which can be operated at temperatures up to 600°C. As a downside, the ceramic material causes a large pressure drop across the filters, and depending on the application their use might not be appropriate. The ceramic filters can be candle filter type, such as the Siemens Westinghouse filter system that has been used at integrated gasification—combined cycle (IGCC) demonstration plants in the United States and Europe. Other examples are the ceramic cross-flow filter and the tube filter used at the Wakamtsu plant in Japan [6].

Several other mechanical methods for particulate or tar removal have been adapted over the course of years for syngas cleaning, such as cyclones, bag filters, baffle filters, wet electrostatic precipitators, etc. However, these conventional cleaning systems were developed for nonreacting flue gas containing mostly CO_2, water vapor, and nitrogen. The composition of the syngas produced from gasifiers brings a new set of challenges that are more suited to the chemical process industry rather than environmental or air pollution control industries. For example, it is desired to operate a syngas cleaning system at the highest possible temperature to avoid excessive energy loss and efficiency reductions. Yet the high temperature presents a risk of undesired reactions between the syngas chemical species (CO, H_2, H_2O, CH_4) and infiltration of air resulting in possible explosion.

6.2.1 Use of syngas for power and heat

It is recognized that gasification has the potential to be more efficient compared with other thermal conversion systems such as combustion and pyrolysis. That is dependent on two major considerations. First the gasification needs to be complete leaving no residual carbon in the ash. The second requirement is that the sensible heat associated with the production of the syngas is preserved prior to entering into a prime mover, provided the chemical energy from the solid and the sensible energy developed during the reaction are completely preserved, which in practice will never be completely realized, then the choice of prime mover dictates the ultimate efficiency of the process. Introducing the syngas or product gas from gasification into GTs is the best possible solution for power generation, because GTs have higher conversion efficiencies than steam cycles. The 7.FA GT from general electric (GE) single cycle net efficiencies reach 39.8%, while the Siemens V94.3 and SGT5-2000E reported to reach net efficiencies of 45% and 43.2%, respectively for IGCC applications in Europe.

The configuration of a GT requires that the turbine blades operate at high temperature; therefore they need to be protected against corrosion. This is one of the peripheral reasons the syngas needs to be cleaned to a minimum set of specifications [7]:

* tars lower than 10 mg/Nm^3
* particulate—2.4 mg/Nm^3
* metals—$0.025-0.1$ ppmw
* $H_2S = 20$ ppmv
* calorific value of minimum 4 MJ/Nm^3

A more forgiving and less expense prime mover is an internal combustion engine (ICE); however, the efficiencies typically are lower compared to GT operation. GTs' efficiencies average around 45%, while the net efficiency for the Jenbacher ICE for IGCC applications is near 37%.

The forgiving nature of the ICE arises from the configuration and construction. Those engines contain the combustion cycle in a large block (the engine block), i.e., water or air cooled. That cooling allows for significant fluctuations in the combustion temperature (between 300 and 500°C) yet maintains a relatively constant output. Conversely that same cooling mechanism lowers the potential work that can be extracted from the combustion, thus lowering the energy.

In the case of using the syngas in ICEs the requirements for tar content ($100 \ mg/Nm^3$) and particulate ($2.4 \ mg/Nm^3$) [7] can be less stringent than in the GT case, while the H_2S, metals, and low heating value (LHV) requirements are similar. Potential efficiencies that could be achieved with an ICE range from 35% to 45%, yet modifications in the injection system are required for many of the engines. The primary design of ICE is for gasoline and a diesel fuel, thus the lower energy dense syngas causes engines to be rated to lower efficiencies. Syngas mixed with biogas from anaerobic digestion can also be a source of fuel for ICE [8].

In terms of particulate the desired concentration produced by the gasifier should be lower than $2-6 \ g/m^3$ [2], which further are removed from the syngas before being fed to the ICE. This is generally achieved due to the lower gas, compared to combustion systems, through the reactor incurring less particulate matter (PM) carryover into the downstream air pollution equipment. Higher concentrations would put more load on the filters and eventually clog the engine.

One benefit to gasification is the production of a lower volume of gas, which allows for smaller gas cleanup equipment, and the ability to retain some of the impurities (metals and minerals) in the solid phase. Some compounds, such as sulfur and chlorine, will still be converted to gas phase products and will need to be removed but other metals and minerals will remain as solid ash and hence are easy to separate. Some waste products, for example plastics, have high chlorine content, whereas others, such as scrap tires, have low sulfur and chlorine content. Therefore the pollution control systems need to be designed based on the feedstock and operating conditions for a given process. A report published in 2009 by the University of California, Riverside (REF), investigated the emissions of many different gasification facilities, including fluidized bed, plasma arc, pyrolysis, and high-temperature gasification systems. Since there is generally little or no oxygen in the flue gas in gasification or pyrolysis systems, the production of dioxins and furans is significantly reduced compared to combustion systems. The emissions reported from

Table 6.1 Emissions from gasification systems

Compound	Approx. range	Units
PM	1—18	mg/Nm^3@7% O_2
HCl	<5—55	mg/Nm^3@7% O_2
NO$_x$	<10—255	mg/Nm^3@7% O_2
SO$_x$	<1—50	mg/Nm^3@7% O_2
Hg	<0.008	mg/Nm^3@7% O_2
Dioxins/furans	~0—0.1	ng TEQ/Nm^3@7% O_2

various gasification systems are shown in Table 6.1. The major economic barrier to gasification is the cost of cleaning the syngas to remove acid and small particles, so the gas can be used with a turbine-powered electric generators or as a chemical feedstock.

New developments in fuel cell technology have been made. Fuel cells convert a mixture of H_2 and CH_4 into energy with efficiency around 40%, and research groups have looked into using syngas as the in-feed. The fuel cells technologies most fitted for syngas use are molten carbonate fuels cells (MCFCs) and solid oxide fuel cells (SOFCs) [7], because they are not poisoned by CO and CO_2. The MCFCs use a nickel-based catalyst and therefore require low H_2S levels in the gas, below 10 ppmv, but present great flexibility in terms of the CO or CO_2 content of the in-feed gas. SOFCs have higher efficiencies due to the high operating temperatures 700—1000°C, but require even lower H_2S concentrations, of 10 ppmv or less. Both types of fuel cells require tar and particulate removal. A study [5] about the SOFC in a combined heat and power system developed by Westinghouse showed a 46% electrical efficiency and 27.5% thermal efficiency to the district heating grid. The same study mentions that the electrical efficiency of MCFC is 50% and 36%—40% for an integrated biomass gasifier with a MCFC system.

The fuel cell technology using syngas is still in the development phase and costs two to three times the price of GT and ICE, but holds a promising market potential for near future. Use of syngas in GT or ICE has made progress over the years, but is still a great challenge to clean the gas to the quality requirements of the engines to maintain an economically viable operation. The syngas is usually cooled down to prepare it for downstream cleaning units. This slightly increases its energy density yet it is still not comparable to the 38 MJ/Nm^3 of natural gas. In reality many of the demonstration and large scale gasifiers burn the syngas in a boiler to produce steam with Energos being one of the most prevalent. It should be noted that the recent financial troubles involving Energos are related to cash flow and contract issues, not the technology [9]. This does not require gas pretreatment and tars are being burned off in the combustor. There are other advantages of this approach, as compared to direct combustion of the feedstock:

- less excess air needed to combust the syngas
- easier to control combustion in the homogeneous phase
- lower volumetric flow rate of gases produced, hence lower cost for the control pollution equipment

The most efficient means for power generation is the IGCC and has been mostly used for coal power plants. The syngas is combusted in a turbine to produce electricity, and the exhaust heat is recovered in a heat recovery steam generator. The steam produced is used in a steam turbine to produce additional electricity, thus increasing the overall combined cycle efficiency. The IGCC usually handles high amounts of feedstock and utilizes GTs, which range from 40 to 200 MW of power, or a combination of engines that can run in a combined cycle.

6.2.2 Tars

The most problematic component in the syngas is the tar, which condenses and clogs piping, valves and can easily progress downstream and foul turbine blades if not removed in the previous steps. With over more than 200 chemical components identified in tars, great scientific efforts were put into designing gasifiers, which would yield <1 g/m^3 of tar [2] in the product gas. Tar condensation would not only damage GT and engines but also deactivate sulfur removal systems, erode compressors, heat exchangers, or ceramic filter [10].

The characteristics of the resulting tars from the pyrolysis stage depend on the operating temperatures and heating rates of the process. Since tars are difficult to remove from the syngas, methods, such as end of pipe tar cleaning, in bed catalytic reforming, or in bed thermal cracking, were developed and continue to be a major interest in the field in order to achieve higher conversion efficiencies at lower costs. The tars decomposition step usually occurs in a separate cracking reactor downstream of the gasifier or using the fluid bed or the freeboard of the reactor itself [11]. This step adds to the overall costs.

Without the use of catalysts, temperature and the addition of coreactants such as steam and oxygen have the main influence on the tar components yielded. Higher temperatures increase the heavy poly-aromatic hydrocarbons concentrations, while less heterocyclic components such as phenols, pyridine, and cresol are generated [12]. The downside of high operating temperatures is that they can have negative effects on heat exchangers and refractory surfaces due to ash sintering in the reactor [10]. Pressure also plays a role in tar production, and above 35 MPa higher conversion rates into gas were observed.

Tar removal via condensation, filtration followed by gas/liquid mixture separation is another possible path to process the syngas and clean it. Such a system includes cyclones, cooling towers, venturi, baghouses, electrostatic precipitators, and wet/dry scrubbers [10]. At the end of this process tars are transferred into wastewater, which requires additional treatment to be disposed of safely.

Alternative tar removal processes include ceramic, metallic, or fabric filters, which many of the times would become clogged because of the "sticky" nature of the tars. They are maintained during the operation by periodic "blowback" to dislodge the built up ash. This procedure requires very good coordination with the entire operating system and is done using automatic sensors and preprogramed durations. If the ash is not at the slagging transition and it remains as a dry dust, the ceramic filters with the blowback operation last approximately 6 months.

Tar yields may vary with different technology types. For example, fixed-bed, updraft, and downdraft gasifiers have high carbon conversion, due to long residence time and would generate less ash. The updraft process generates the highest amount of tars in the product gas.

6.3 TECHNOLOGY CHALLENGES

The type and configuration of the technology used influences the product gas quality significantly. Also, based on the feedstock characteristics and the targeted end utilization of the products, the best approach is chosen. Well-developed gasification technologies present challenges at large scale. For example, *the updraft technology* yields high tar content syngas, $10-20$ wt% and requires further treatment before other uses. *Downdraft* technology yields more char and requires a dry feedstock, with <20 wt% moisture content. *Fluidized bed gasification* can be applied to various types of feedstock due to the technology flexibility. Dual fluidized bed gasifier is a technology where gasification process heat is supplied indirectly by a heat carrier, usually from combusting the char created in the gasification stage. This presents an advantage over autothermal gasification which mostly uses air to partially combust the feedstock, by avoiding an air separation unit usually required for a nitrogen–free product gas. If for large gasification plants, using pure oxygen to supply the heat for the process is economically viable [13], for medium sized plants, the autothermic technologies might seem more attractive.

The authors of study [14] looked at different gasification technologies and concluded that *bubbling fluidized bed gasification* is one of the most widely demonstrated technology, over a wide range of conditions and using a variety of biomass feedstock. Operating temperatures above 1200°C are favorable for less tar formation and maximized release of H_2 and CO.

Circulating fluidized beds require small material size and generate a product gas with less tar concentrations. However, the high temperatures can cause ash melting–related problems and temperature gradients may occur in the direction of solid flow [14]. The high velocities of fuel particles may lead to equipment erosion.

Fluidized bed technology requires a higher investment and becomes financially more attractive when higher amounts of feedstock are available. For lower amounts of feedstock available, for example MSW from a town or few communities amounting to 25 tons per day, *fixed-bed gasification technology*, is more appropriate. The fixed-bed technology generates less fly ash compared with fluidized bed gasification. This technology also generates more tars but has the ability to handle heterogeneous feedstock much more efficiently than others.

Supercritical water gasification, described in more detail in Chapter 2, Fundamentals of Gasification and Pyrolysis, also presents some challenges. The most important is the salt precipitation, which occurs due to the rapid decrease in the salts solubility under supercritical conditions. This can cause tube plugging and several designs have been developed to avoid this issue [15]:

- Reverse flow reactors that feed the material from the top at supercritical temperatures and the water from the bottom at subcritical conditions.
- Transpiring wall reactor in which water flows through a porous reactor liner, avoiding salt deposition and corrosion on the reactor walls.
- Salt separation unit that uses the same principal as the reverse flow reactor, but serves as a pretreatment vessel.

- Fluidized bed reactor for the supercritical water gasification process.

Most of these technologies are in pilot scale stage only.

Other significant challenges with the supercritical water gasification technology are related to corrosion, heat exchanger efficiency, and biomass feeding. The feedstock needs to be in a slurry pumpable state to be delivered to the reactor, and achieving that state requires energy input and good initial design. Importantly to achieve a high value gas, supercritical water systems are not suitable. As discussed previously, the dryer the feedstock the higher the syngas heating value. Therefore introducing a solid feedstock into a water bath requires significant amounts of energy. It should be noted that supercritical water oxidation systems are best suited for destruction of unwanted chemicals (i.e., chemical and biological weapons) and sterilization processes.

6.3.1 Ash and char challenges

Ash can cause several problems in a gasification system [16]. It forms slag by fusing together and in time it can deposit on the walls of the reactor. The alkali metals in the feedstock react with silica forming silicates (e.g., K_2SiO_3) or with sulfur forming alkali sulfates (e.g., K_2SO_4). The melting temperatures of the newly formed compounds are lower than 700°C [5]. Their melting and tendency to stick to the reactor wall causes large deposits, usually referred to as clinkers. This clinker needs to be removed periodically; otherwise it significantly impacts the heat transfer and overall efficiency of the process. Additionally, as the ash builds up on reactor walls, the gas flow path can be significantly altered creating conversion losses and contributing to erosion problems. Costs are associated to remove clinkers depending on the size of the reactor and the type of waste used, which would generate more or less ash.

Char can also become problematic in a gasification plant if there are no other uses for this by-product. Owing to its high carbon content, char can be used as coal replacement in incineration applications or as feedstock for pyrolysis and gasification.

6.3.2 Other challenges

The exhaust gas volume is known to be lower for the gasification process rather than in conventional WTE combustion. For the complete combustion of feedstock, excess air above stoichiometric requirements is needed, while gasification is carried out at equivalence ratios (air/fuel ratio) between 0.25 and <1.0, depending on the feedstock and technology. In reality, if the product gas is subsequently combusted in a steam boiler, the amount of flue gas sent to pollution control equipment can be comparable to WTE technology and costs for this step is not significantly reduced as some gasification developers claim. The volume of gas produced per ton of gasified waste was found to be around 6000–7000 Nm^3/t [10].

It has been demonstrated through multiple attempts by numerous technology companies that many of the pyrolysis or gasification facilities around the world were shut-down or did not even make it past the demonstration scale point. The issues with every one of them are facility specific, but studies [1] have been carried out to determine a common ground behind these failures.

One key technical issue was using a large scale up factor, and thus the results from the pilot scale or prototype facilities were not valid anymore on larger scale. Some of the facilities faced technical issues after being build based on theoretical design and integrated components from other operating facilities. Generally the unit process scale up factor should not be more than one-third larger than the operating facility [1].

Many gasification experiments are conducted in batch reactors, which are not suitable for industrial applications, where continuous processes are in demand. However, continuous experimental approaches face significant challenges, which need to be overcome to scale up the process and be applicable in industry.

Another well-known issue extensively discussed within the scientific community is the lower emissions coming out of a gasification process compared to combustion. In reality there are several different aspects when considering the emissions from a facility, such as presence of pollutants in the incoming feed or air pollution control methods applied [17].

Stoller and Niessen [1] have studied RDF pyrolysis facilities where higher dioxin/furan formation was encountered because of unreacted char presence in the gas stream. MSW or RDF facilities could also face a contamination problem, which downstream would release higher emission concentration than mass burn system, and increase the costs associated with pollution control equipment and chemicals required to comply with the regulations.

Contamination of MSW also contributes to slagging and tube fouling, which eventually leads to high boiler maintenance, outages, and lower heat recovery efficiencies. As explained in the "Feedstock" subchapter, MSW is a problematic type of feedstock, which requires extensive study and knowledge before designing a MSW gasification facility. Cross contamination, a general problem in these types of facilities, can be avoided by placing a material recovery facility (MRF) at the front-end of the gasification system.

In addition to contamination and handling issues, MSW characteristics also vary with community and season. Therefore the success of a pyrolysis/gasification facility also relies on good knowledge about your feedstock characteristics, corrosive compounds, energy value, and eventual changes with season. Serious efforts need to be put into cost prediction, emphasizing the estimated revenues and possible changes in the market, which would change the projected numbers. This guideline is applicable for any type of feedstock.

6.4 FEEDSTOCK CHALLENGES

Feedstock availability and market price often decide the success and profitability of a technology even if the developer receives some sort of financial incentives from the government.

The discussion in this section focuses on the two most common types of feedstock used in the gasification process: MSW and biomass.

6.4.1 Municipal solid waste

The heterogeneous character of the waste, along with its moisture content, is among the two most important challenges gasification plant operators face. The waste requires certain preparation, to make it more homogeneous and obtain optimal heat and mass transfer rates within the reactor. Homogenizing different types of materials, such as fluff (shredded light fraction, dust, soft fibers accumulated in clumps), plastics of all sort, paper, and removing unwanted materials could be challenging and will definitely add to the capital and operation cost yet increases the homogeneity and heating value of the feedstock.

Many MSW or RDF gasification plants had material handling problems, which lead to their shut-down, because of high costs associated with lengthy startup and outages. Some of these names include the Baltimore, MD Monsanto Landgard pyrolysis facility or the Torrax Systems, Inc. pyrolysis process [1]. Challenges in designing the handling, sorting, and separation stage of the process are related to the inconsistent character of the waste, which varies significantly with season and geographically. A design, which works in one part of the world, most likely would need in depth analysis before it can become successful in other parts.

One of the most common types of challenges in the field of material handling, especially when dealing with RDF, is the high chance of explosion in the shredders. Other known issues with MSW and RDF handling and preprocessing [1] upstream of gasification plants are:

- the material binding to itself in storage facilities and even in reactors,
- shredded MSW over time deteriorate the feed conveyors due to its abrasive character,
- the feeding step can also present challenges, many times blockages can happen and restarts are needed, and
- the whole pretreatment step needs energy input, which needs to be considered in the overall energy balance.

If no pretreatment stages are used, other problems may occur downstream, which many times can be more significant on long-term operation:

- damage of reactor parts done by bulky materials
- corrosion caused by unremoved inorganics or materials with high PVC, chlorine content (such as cables)
- higher dioxins emissions caused by heavy metals or batteries
- more char and ash generated

Thus the gasification of MSW should be implemented only for post-recycling feedstock, especially removing materials such as inorganics and metals that have little to no heating value. Even if plasma gasification is the chosen technology, known to treat a variety of waste, the long-term success, higher efficiencies, and better economics are dictated by the pretreatment step.

The challenge of pretreatment stage (MRF) coupled with gasification design is balancing the amount of separation with the quantity and quality of fuel sent into the reactor. The system should be flexible enough to accommodate possible changes in the market price, without large investment and long outages needed for modifications.

The equipment used in the preprocessing stage involves conveyors, shredders, trommel screens, vibrating screens, air classifiers, and magnetic- and eddy-current separators, which increase the project capital cost by tens of millions of dollars. To that, maintenance and operation costs are added. If a drying stage is required, the cost and energy input increase significantly.

Therefore an operator will need to find a balance and determine the cost-effective way to preprocess the waste enough to maintain an efficient and stable gasification process, while getting some revenue from the materials separated. Most of the US plant operators follow the Institute of Scrap Recycling Industries (ISRI) Scrap Specifications Circular that defines the different grades of metal (ferrous, nonferrous), glass cutlet, paper, electronics, and tire scrap. These specifications and guideline are accepted and used by selling and buying industries to trade these commodities. A gasification plant operator might choose not to separate from the feeding stream some lower grades of paper, such as Old Newspaper grade, which would not be valuable on the commodity trading market.

Another good example by Dodge [3] is tires, which are a valuable source of fuel, but are embedded with steel threads. When tires will be gasified, a detailed analysis is necessary before deciding the level of preprocessing needed, which determines the amount of steel which can be removed and the costs associate with that. Parts of steel embedded in the tire are acceptable for use in a plasma gasification plant for example, and a lower level of processing would make more sense in this case.

6.4.2 Biomass

The biomass composition influences the product gas. It was found that higher yields of gas are released from biomass with increased H/C ratio [2], while higher oxygen content in the biomass requires lower equivalence ratios because of the oxygen available for gasification.

Nevertheless when analyzing woody biomass, the composition of the syngas was not found to vary widely, with only small differences found in literature. For example, straw would yield a gas with higher hydrogen content, while soft woods were found to generate the highest LHV syngas [7]. Still the conditions of the process were found to have a much greater influence on the quality and quantity of the syngas.

Often times the biomass gasified under supercritical conditions requires pretreatment for particle size reduction in order to avoid pumpability problems created by clogging. Sludge containing 40% dry matter was found optimum to be successfully pumped [15].

Particle size reduction also increases the reactive surface condition, important for all gasification technologies for better heat and mass transfer and lower residence time requirements. Fluidization also adds to better intermixing and fuel conversion [18], and higher heat and mass transfer is achieved in fluidized bed instead of moving bed gasification. In some cases the reduced particle size requires not only lower residence time but also lower operating temperatures, improving the overall process efficiency.

6.5 CORROSION

Corrosion is among the most important problems in industrial applications. Corrosion can occur on the reactor walls or in the heat recovery steam boilers, causing great challenges for plant reliability. Heat exchangers used for plasma gasification technology, at temperatures around 1500°C face great challenges due to corrosion. High temperatures strain the steel and leave the material unprotected against corrosive gases sometimes evolved during gasification. The degree of acidic components in the gas depends mainly on the feedstock gasified and it can be prevented with the use of high corrosion restive materials in the high temperatures zones. This is mainly a challenge when MSWs are treated, and use of steel alternative materials would increase the initial costs of the plant.

The most aggressive type of corrosion was found to be in the superheater, where hydrogen chloride and metal chlorides can accelerate the high-temperature corrosion of boiler surfaces, especially superheater tubes [19]. The presence of metal chlorides in MSW plants can be easily detected in fly ash and boiler deposits. A PhD study by Sharobem [19] from Colombia University found that the corrosion behavior under chloride salts was much more severe than under HCl atmosphere.

The same study investigated the corrosion behavior of metal sulfates and found that sulfates salts shown semiprotective of boiler tube surfaces at temperatures up to 550°C. The corrosion layer formed below sodium sulfate was an order of magnitude lower than under sodium chloride [19].

Several approaches to prevent corrosion [15] have been proposed:
- Circulating flow reactors
- Use of high corrosion resistant materials
- Reducing the operating temperatures to 400°C, the optimum to prevent corrosion

Reducing the operating temperature might be one successful way of going to prevent corrosion, possible when using liners and coatings to decrease the heat transfer between surfaces. The temperatures, however, affect greatly the carbon conversion rates in the gasifier and in order to achieve the same efficiencies catalyst need to be used.

Another possible solution to mitigate corrosion, proposed by Sharobem [19], was the addition of corrosion additives such as sulfur containing chemicals: ChlorOut ammonium sulfate—based additive by Vattenall or the sulfur recirculation technology by Gotaverken Mijo. Overall the same balance between cost and efficiency must be maintained for successful plant design and further operation.

6.6 AUXILIARY LOADS AND ENERGY EFFICIENCY

The stages of transportation, preprocessing, and sometimes production of feedstock are highly energy consuming processes, which need to me accounted for in the life cycle assessment of a gasification technology.

For example, biomass grows accounts for 50% of the energy in the integrated gasification cycle, while pretreatment and transportation consume 43% and 4%—16% energy, respectively [20].

On today's market, there are several technological options, but unfortunately many of them are not yet matured. Use in fuel cells, methane, or hydrogen generation, and production of some chemicals still need further development. At this time the most successful option of using product gas from a gasification process is combustion in a boiler coupled with steam or Organic Rankine Cycle (The Organic Rankine Cycle is working on the same principle as the Rankine Cycle; however instead of water, it uses an organic, high molecular mass fluid as working fluid). The efficiency of a conventional system of 28%—31% could be increased up to 37%—41% with alternative engines and 50% with GTs in combined cycles [10]. The challenge remains cleaning the syngas from tars, particles, and other pollutants to the standards required by ICE and GTs, in an efficient and cost-effective manner.

Panepinto et al. [10] discussed the higher internal parasitic loads associated with alternative technologies compared to conventional incineration, and to what they might be attributed:

- Preprocessing, such as shredding and drying
- Use of pure oxygen as gasifying agent
- Higher temperatures required by gasification systems, such as AlterNRG, Ebara, Nippon Steel
- Syngas cleaning

Table 6.2 Electrical and thermal efficiency for MSW combustion and gasification process

	MSW combustion	MSW gasification—ICE	MSW gasification—GT
Thermal conversion efficiency (%)	80	70	70
Power generation efficiency (%)	31	38	30

Source: Adapted from D. Panepinto, V. Tedesco, E. Brizio, G. Genon, Environmental performances and energy efficiency for MSW gasification treatment, Waste Biomass Valoriz. 6 (1) (2014) 123−135.

Thermal efficiency also depends on the initial air temperature supplied to the gasifier. High-temperature air gasification (HTAG) was found to be a more efficient process [21] which generates a product gas with higher heating value. Pre-heating the air or steam is also an energy consuming process, so ways to capture some of the waste heat from an engine for example, could increase the overall efficiency.

Table 6.2 shows the thermal and power efficiencies for MSW combustion compared to MSW gasification and use in an ICE and GT, respectively.

Steam turbines in waste burn plants have a typical power efficiency of 31%, while gas engines and GTs reach 34%−41% and 30%, respectively.

Prabowo et al. [22] looked at CO_2 gasification and compared the thermal efficiency of the system to that of using steam as gasifying agent. It is well known that steam gasification requires high inputs of energy because of the latent heat necessary to convert water into steam. Steam gasification turned out to be less efficient than pyrolysis at high operating temperatures of 950°C. CO_2 as gasification agent decreased the thermal efficiency of the gasifier at temperatures of 750°C and increased it above that, achieving the maximum at 850°C.

REFERENCES

[1] P.J. Stoller, W.R. Niessen, NAWTEC 17-2348, pp. 1−15, 2009.
[2] A. Malik, S.K. Mohapatra, Biomass-based gasifiers for internal combustion (Ic) engines—a review, Sadhana 38 (2013) 461−476.
[3] E. Dodge, Plasma-gasification of waste: clean production of renewable fuels through the vaporization of garbage, Cornell University, Johnson Graduate School of Management, Queens University School of Business, July 2008.
[4] B. Messenger, rePlanet closes 191 recycling centres in California, Waste Management World, 2016.

[5] L. Wang, C.L. Weller, D.D. Jones, M.A. Hanna, Contemporary issues in thermal gasification of biomass and its application to electricity and fuel production, Biomass Bioenergy 32 (7) (2008) 573−581.

[6] R. Zevenhoven, P. Kilpinen, Control of Pollutants in Flue Gases and Fuel Gases, Helsinki University of Technology, Espoo, Finland, 2001.

[7] A. Molino, S. Chianese, D. Musmarra, Biomass gasification technology: the state of the art overview, J. Energy Chem. 25 (1) (2016) 10−25.

[8] C. Marculescu, V. Cenusa, F. Alexe, Analysis of biomass and waste gasification lean syngases combustion for power generation using spark ignition engines, Waste Manag. 47 (2016) 133−140.

[9] L. Walsh, Energos calls in administrators, Ends Waste & Bioenergy, 2016. [Online]. Available from: http://www.endswasteandbioenergy.com/article/1402205/energos-calls-administrators (accessed: 01.03.17).

[10] D. Panepinto, V. Tedesco, E. Brizio, G. Genon, Environmental performances and energy efficiency for MSW gasification treatment, Waste Biomass Valoriz. 6 (1) (2014) 123−135.

[11] B.J. Vreugdenhil, R.W.R. Zwart, Tar formation in pyrolysis and gasification, Energy Research Center of the Netherlands, ECN-E–08-087, June 2009.

[12] S. Kumar, S. Suresh, S. Arisutha, Production of renewable natural gas from waste biomass, J. Inst. Eng. Ser. E 94 (2013) 55−59.

[13] C. Pfeifer, S. Koppatz, H. Hofbauer, Steam gasification of various feedstocks at a dual fluidised bed gasifier: impacts of operation conditions and bed materials, Biomass Convers. Biorefinery 1 (1) (2011) 39−53.

[14] J.P. Ciferno, J.J. Marano, Benchmarking Biomass Gasification Technologies for Fuels, Chemicals and Hydrogen Production, U.S. Department of Energy National Energy, 2002.

[15] O. Yakaboylu, J. Harinck, K.G. Smit, W. De Jong, Supercritical water gasification of biomass: a literature and technology overview, Energies 8 (2) (2015) 859−894.

[16] G. Ionescu, E.C. Rada, Material and energy recovery in a municipal solid waste system: practical applicability, Int. J. Environ. Resour 1 (1) (2012) 26−30.

[17] G. Ionescu, D. Zardi, W. Tirler, E.C. Rada, M. Ragazzi, A critical analysis of emissions and atmospheric dispersion of pollutants from plants for the treatment of residual municipal solid waste, UPB Sci. Bull. Ser. D: Mech. Eng. 74 (4) (2012) 227−240.

[18] A. Pollex, A. Ortwein, M. Kaltschmitt, Thermo-chemical conversion of solid biofuels, Biomass Convers. Biorefinery 2 (1) (2012) 21−39.

[19] T.T. Sharobem, Mitigation of High Temperature Corrosion in Waste-to-Energy Power Plants, Columbia University, New York, NY, 2017.

[20] A. Sanna, Advanced biofuels from thermochemical processing of sustainable biomass in Europe, Bioenergy Res. 7 (1) (2014) 36−47.

[21] C. Lucas, D. Szewczyk, W. Blasiak, S. Mochida, High-temperature air and steam gasification of densified biofuels, Biomass Bioenergy 27 (6) (2004) 563−575.

[22] B. Prabowo, K. Umeki, M. Yan, M.R. Nakamura, M.J. Castaldi, K. Yoshikawa, CO_2-steam mixture for direct and indirect gasification of rice straw in a downdraft gasifier: laboratory-scale experiments and performance prediction, Appl. Energy 113 (2014) 670−679.

CHAPTER SEVEN

Economic Summary

Contents

7.1 INTRODUCTION

The economics of waste thermal treatment systems in the United States (U.S.) currently are not at a state that makes them competitive with landfilling. The high capital cost of these facilities combined with the competitive natural gas electricity markets makes the economic viability of waste gasification and pyrolysis in the U.S. difficult to attain. However, as landfill capacity is used up and waste has to be transported further distances, the margin of cost between landfilling and energy recovery is starting to decrease. In Europe and Asia, energy recovery systems are more prevalent due to policy initiatives that enforce diversion from landfills and provide incentives for waste thermal treatment systems as part of renewable energy initiatives.

This chapter discusses the general economics of waste gasification and pyrolysis technologies and includes cost breakdowns of a few waste gasification and pyrolysis facilities provided by the companies and independently analyzed by the Earth Engineering Center at City College of New York (EEC | CCNY). The siting process for facilities is also discussed as well as the policies that have been implemented abroad that have made waste gasification and pyrolysis more economically viable. Finally, the economics of energy recovery is compared to that of other waste practices

Gasification of Waste Materials.
DOI: http://dx.doi.org/10.1016/B978-0-12-812716-2.00007-8

in the waste management hierarchy, specifically recycling, composting, and landfilling.

7.2 COST BREAKDOWN OF WASTE GASIFICATION AND PYROLYSIS PLANTS

The economic benefit of a waste gasification or pyrolysis facility is quantified by its net revenue. This is defined as the difference between the total income and total costs to the facility. The total income for a waste thermal treatment facility is the revenue from the gate fee and from the final product sold which can be electricity, heat, hydrogen, or liquid fuels and chemicals depending on the process. The gate fee is what is charged to the waste provider, such as a municipality, to accept the waste and is dictated by the costs of the facility since it must be adjusted in order to generate enough income for the facility to continue operations. The gate fees and energy revenues are variable income because they are dependent on the quantity of the waste being treated and the quantity of energy being generated. The fixed costs of a waste gasification or pyrolysis facility are the capital cost and the fixed taxes and permit fees. The variable costs are for personnel, operation and maintenance (O&M), utilities, and materials [1].

Waste gasification and pyrolysis technologies are more expensive than waste-to-energy (WtE) technologies because of higher capital costs resulting from the added expenses of feedstock pretreatment and gas cleaning and conditioning stages (in the case that the syngas is not combusted) that are required. Furthermore, gasification and pyrolysis are not as feedstock flexible as WtE, because slight variation in the waste feedstock could potentially produce off-spec product or compromise the system equipment thus reducing the reliability of the technology from an investment standpoint. Therefore, the challenge in the development of waste gasification and pyrolysis technologies at commercial scale is that they are being considered a risky investment compared to WtE, because they are not yet proven technologies with years of commercial operations to prove their reliability. However, this is a chicken and egg debacle because waste gasification and pyrolysis will not be proven technologies unless a risk is taken and investments are made to develop the technologies at commercial scale. As municipalities develop and innovate their waste management infrastructure in order to meet their zero waste targets, there is a growing interest in the potential incorporation of waste gasification and pyrolysis

Table 7.1 Cost breakdown of energy recovery facilities

Technology	WTE[a]	WTE[b]	Plasma gasification	Plastics pyrolysis[a]	Plastics pyrolysis[b]
Capacity (tpd, tons per day)	2800	3000	750	36	24
Capital cost ($/ton)	N.A.[a]	N.A.	807	1195	400−700
Operating cost ($/ton)	27	30	42	135	30−35[b]
Total cost ($/ton)	N.A.	N.A.	123	546	N.A.
Total revenue ($/ton)	19	22	130	95	N.A.
Gate fee ($/ton)	49	49	65	50	N.A.

[a]N.A. indicates that this information was not provided by the company.
[b]Unit is in $/barrel.

as part of their waste management strategy. In addition, federal and local incentives in regions such as Europe are allowing this risk taking to occur with a safety net being provided by the government.

7.2.1 Waste gasification and pyrolysis economics: case studies

Data availability on the economics of waste gasification and pyrolysis facilities is sparse in the public domain. Table 7.1 presents cost breakdowns of WtE, gasification, and pyrolysis facilities that were provided by the companies and independently analyzed by EEC|CCNY. The purpose of this table is to provide a sample of cost breakdowns, but it should be noted that these do not necessarily represent typical cost breakdowns for the technologies. Costs will vary based on factors including design, feedstock flexibility, reliability, and incentives.

7.3 COMMERCIALIZATION OF WASTE GASIFICATION AND PYROLYSIS FACILITIES

Ownership of waste gasification and pyrolysis facilities can be public, private, or public—private. Public ownership is typically through a municipal government, authority, or agency, and private ownership is through a private corporation, partnership, or sole proprietorship. Public ownership of highly capitalized waste facilities is recommended, because it requires less time to finance and implement, and it is easier to obtain tax-exempt debt financing [2]. All waste gasification and pyrolysis facilities undergo six main stages in their commercialization: procurement, planning, permitting, construction, commissioning, and operation.

There are four major contractual arrangement options that can be pursued by the private sector provider and waste management authority in the procurement of a waste gasification or pyrolysis facility [3]. These options are summarized in Table 7.2.

Planning is submitted to the appropriate regulatory agencies and the issues that need to be addressed in the submission are facility siting, traffic, air emissions and health effects, dust and odor, flies, vermin and birds, noise, litter, water resources, visual intrusion, size and landtake, and public concerns. In terms of facility siting, it is generally recommended that the facilities are built in areas where land was previously allocated for this type of industrial activity and where there is a good transport infrastructure. It is

Table 7.2 Contractual arrangement options for energy recovery facilities

Number of separate parties involved	Contractual arrangement description
4	*Separate design; build; operate; and finance*: Waste authority contracts separately for the works and services needed and provides funding by raising capital for each of the main contracts. The contract to build the facility would be based on the council's design and specification and the council would own the facility once constructed
3	*Design and build; operate; finance*: The private sector provides both the design and construction of a facility to specified performance requirements. The waste authority owns the facility that is constructed and makes separate arrangements to raise capital. Operation would be arranged through a separate operation and maintenance contract
2	*Design, build, and operate; finance*: The design, build, operation, and maintenance contracts are combined. The waste authority owns the facility once constructed and makes separate arrangements to raise capital
1	*Design, build, finance, and operate*: This contract is a design, build, and operate, but the contractor also provides the financing of the project. The contractor designs, constructs, and operates the plant to agreed performance requirements. Regular performance payments are made over a fixed term to recover capital and financing costs, operating and maintenance expenses, plus a reasonable return. At the end of the contract the facility is usually transferred back to the client in a specified condition

Source: Data from DEFRA UK (2002).

advantageous if the facility is located on site of where the feedstock will be provided, such as refuse-derived fuel (RDF) and mechanical biological treatment facilities. Also, the optimized export of the final product, specifically electricity and/or heat, to the host users or to the national grid must be considered in the planning as well. Finally, traffic layouts must also be submitted to limit impact of large heavy goods vehicles on roads surrounding the facility and nearby residential areas [3].

Planning must address the environmental and health impacts of the proposed facility. Air pollution control systems must be implemented to meet limits of pollutants in flue gases as outlined in the environmental regulatory framework of the region; for the United States (U.S.), this is the National Ambient Air Quality Standards (NAAQS) and for Europe, it is the Industrial Emissions Directive (IED). Emissions of the facilities need to be monitored continuously and for emissions, such as for dioxins and furans, for which continuous monitoring is not possible, periodic sampling and measurements are required. Dust, odor, and vermin control are additional aspects that need to be minimized in the planning and construction of these facilities to prevent negative impacts on surrounding residential areas. Finally, good facility design in terms of functionality and aesthetic quality must also be addressed [3].

Permitting involves applying for environmental permits from the appropriate regulatory agencies in order to commercially operate. Permits ensure on-going compliance with regulatory requirements and use of best available techniques. The scope of the project proposal along with the local environmental circumstances will determine the nature and complexity of the permit and consequently how long it will take for the permit to be granted [3]. Permitting is one of the longest stages in the implementation of thermal treatment facilities and frequently can delay projects and even force them to close down either because of financial reasons or because of the efforts of public opposition groups that use the permitting period to fight against the construction of the facilities.

Construction, commissioning, and operations are the last stages of implementation of a commercial facility. Construction of facilities generally takes 1—2 years depending on the size and design of the facility. Commissioning is the period of operation at commercial scale that operators use as a test run to work out any "kinks" in the system and to prove continuous reliable operation. Usually commissioning involves test campaigns that run different feedstocks through the system for a specified period of time to prove fuel flexibility, availability, and reliability of the technology at commercial scale.

7.4 IMPACT OF POLICY ON WASTE THERMAL CONVERSION ECONOMICS

The economics of waste thermal conversion is significantly impacted by policy and waste legislation. In Japan, there is a prevalence of waste thermal conversion, because the country does not have enough land space for landfilling. Furthermore legislation dictates that waste generated within a prefecture must be treated within that prefecture; therefore modular small-scale thermal conversion facilities are needed which makes it more economical in terms of capital cost as well as reduced O&M costs, because the technologies are more reliable and proven at low capacities. In Europe the European Union (EU) Landfill Directive requires the diversion of biodegradable municipal waste from landfill with total biodegradable MSW sent to landfill being no more than 35% by 2016. As a result of this directive, there has been an increase in waste thermal conversion in Europe [4]. Feed-in tariffs and tax credits can further improve the economics of waste thermal conversion technologies. In Europe and the U.S., renewable energy tax credits can be provided to waste thermal conversion technologies for the biogenic fraction of MSW that is processed to generate electricity.

Waste thermal conversion is not as prevalent in the U.S. as it is in Europe and Japan due to the fact that it is not cost competitive with landfilling. The landfill tipping fee in the United States is $49.78/ton of MSW [5]. By comparison the gate fees for typical WTE facilities is $100−120/ton of MSW. The margin of cost however is closing in certain regions, where the waste has to be transported farther distances in order to be landfilled. Also, zero waste initiatives have been implemented as part of the waste management campaigns of many municipalities throughout the U.S. If a national landfill directive, similar to the EU's, is implemented, this will most likely expand the waste thermal conversion landscape in the U.S.; however, incentives and tax credits will still be necessary to make this technology economical during its growing period into a proven technology at commercial field scale.

7.5 COST COMPARISON OF WASTE MANAGEMENT PRACTICES

The optimal waste management infrastructure is one that is integrated and incorporates all practices of the waste management hierarchy to the degree that is necessary given the composition of the MSW. Table 7.3

Table 7.3 Typical costs associated with waste management practices

Waste management practice	Material recovery (low mechanical intensity)	Material recovery (high mechanical intensity)	Composting (low end)	Composting (high end)	WTE (mass burn)	WTE (modular)	WTE (RDF)	Landfilling
Waste processed	Recyclables only		Organics only		Nonrecyclable, noncompostable refuse			
Typical capital cost ($/tpd)	10,000–20,000	20,000–40,000	10,000–20,000	25,000–50,000	80,000–120,000	80,000–120,000	80,000–120,000	25,000–40,000
Typical O&M cost ($/ton)	20–40	30–60	20–40	30–50	40–80	40–80	20–40	10–120
Final product	Recycled consumer products	Recycled consumer products	Fertilizer, biogas	Fertilizer, biogas	Electricity	Electricity	Electricity	Biogas if LFGTE[a] is employed

[a]LFGTE, landfill gas to energy.

Source: Data from G. Tchobanoglous, F. Kreith, Handbook of Solid Waste Management, second edition, McGraw-Hill, New York, NY, 2002.

presents the typical capital and operating costs of the waste practices in the waste management hierarchy with WTE technology as the representative technology for energy recovery. It should be noted that the profit opportunity offered by the final product of the different practices is not included in this table but must be taken into account when evaluating the overall economic impact of a waste practice.

Based on the results in Table 7.3, WTE has higher capital costs than recycling and composting; however, WTE of RDF is cost competitive with high mechanical intensity recycling and high end composting in terms of O&M costs. Landfilling is cost competitive recycling and composting in terms of capital cost and has a wide range of O&M costs. Overall landfill costs are projected to increase due to the continuous increase in historic landfill tipping fees. The other waste management practices, including WTE, will become more cost competitive with landfills as nearby landfills reach capacity, and consequently waste has to be transported farther distances for disposal.

7.6 CONCLUDING REMARKS

Currently, thermal conversion of waste is not cost competitive with landfilling due to high capital costs of the facilities. The market for waste gasification and pyrolysis technologies can grow if municipalities push to meet their zero waste initiatives for landfill diversion through increased recycling, composting, and energy recovery. This will consequently produce more homogeneous streams for collection and processing and reduce front-end costs for waste gasification and pyrolysis facilities. Short-term economic support from federal policies and incentives can allow for development in commercialization of these technologies at field scale which in turn could make waste gasification and pyrolysis more economically viable and cost efficient in the long term.

REFERENCES

[1] ISWA, Alternative Waste Conversion Technologies, January 2013.
[2] G. Tchobanoglous, F. Kreith, Handbook of Solid Waste Management, second edition. McGraw-Hill, New York, NY, 2002.
[3] Department for Environment, Food, and Rural Affairs, Advanced Thermal Treatment of Municipal Solid Waste, www.defra.gov.uk. February 2012.
[4] Eurostat, Municipal Waste Statistics, http://ec.europa.eu/eurostat/statistics-explained/index.php/Municipal_waste_statistics, January 2017.
[5] Office of Resource Conservation and Recovery, U.S. EPAK, Historic Tipping Fees and Commodity Values. February 2015.

INDEX

Note: Page numbers followed by "*f*" and "*t*" refer to figures and tables, respectively.

Printed in the United States
By Bookmasters